芒果
种植管理技术（中英文版）

◎丁哲利　周泽雄　陈　丹　主编

中国农业科学技术出版社

图书在版编目（CIP）数据

芒果种植管理技术：汉、英 / 丁哲利，周泽雄，陈丹主编 . -- 北京：中国农业科学技术出版社，2022.11
ISBN 978 - 7 - 5116 - 6042 - 8

Ⅰ.①芒…　Ⅱ.①丁…②周…③陈…　Ⅲ.①芒果—果树园艺—汉、英　Ⅳ.① S667.7

中国版本图书馆 CIP 数据核字（2022）第 225127 号

责任编辑　周丽丽
责任校对　王　彦
责任印制　姜义伟　王思文

出 版 者	中国农业科学技术出版社
	北京市中关村南大街 12 号　邮编：100081
电　　话	（010）82109194（编辑室）　（010）82109702（发行部）
	（010）82109702（读者服务部）
网　　址	https:// castp.caas.cn
经 销 者	各地新华书店
印 刷 者	北京建宏印刷有限公司
开　　本	185 mm×260 mm　1/16
印　　张	11.25
字　　数	370 千字
版　　次	2022 年 11 月第 1 版　2022 年 11 月第 1 次印刷
定　　价	108.00 元

版权所有·侵权必究

本书由海南省农业对外交流合作中心2021年部门预算项目"农产品促销与农业交流合作""中国—柬埔寨热带技术培训班"项目资助。

本书部分工作获中央级公益性科研院所基本科研业务费专项"热区（香蕉、芒果、火龙果）土壤酸化调控原理与技术集成示范"（1630092022001）、国家重点研发计划子课题"芒果菠萝周年供应与花果调控技术集成与示范"（2020YFD1000604-X）项目资助。

特此致谢！

《芒果种植管理技术（中英文版）》
编委会

主　　编：丁哲利　　周泽雄　　陈　丹
副 主 编：陈　媛　　薛晶洁　　周兆禧　　何应对
翻　　译：薛晶洁　　张荣敏

编　　委：丁哲利　　中国热带农业科学院海口实验站
　　　　　周泽雄　　海南省农业对外交流合作中心
　　　　　陈　丹　　海南省农业对外交流合作中心
　　　　　陈　媛　　海南省农业对外交流合作中心
　　　　　薛晶洁　　海南省农业对外交流合作中心
　　　　　张红亮　　海南省农业对外交流合作中心
　　　　　蒙忠武　　海南省农业对外交流合作中心
　　　　　周兆禧　　中国热带农业科学院海口实验站
　　　　　何应对　　中国热带农业科学院海口实验站
　　　　　明建鸿　　中国热带农业科学院海口实验站
　　　　　田　昌　　湖南农业大学资源环境学院
　　　　　刘　娣　　九江学院资源环境学院
　　　　　明斯妤　　中国热带农业科学院热带作物品种资源研究所

Practice Scheme of Mango Plantation

(Chinese and English version)

Editors-in-Chief: DING Zheli, ZHOU Zexiong, CHEN Dan

Associate Editors-in-Chief: CHEN Yuan, XUE Jingjie, ZHOU Zhaoxi,

HE Yingdui

Translator: XUE Jingjie, ZHANG Rongmin

Editor Boards:

Hainan Agricultural Foreign Exchange and Cooperation Center

    ZHOU Zexiong, CHEN Dan, CHEN Yuan, XUE Jingjie,

    ZHANG Hongliang, MENG Zhongwu

Haikou Experimental Station, Chinese Academy of Tropical Agricultural Sciences

    DING Zheli, ZHOU Zhaoxi, HE Yingdui, MING Jianhong

College of Resources and Environment, Hunan Agricultural University

    TIAN Chang

School of Resources and Environment, Jiujiang University

    LIU Di

Tropical Crops Genetic Resources Institute, Chinese Academy of Tropical Agricultural Sciences

    MING Siyu

目 录

1 绪 论 ·· 1
　1.1 芒果概述 ··· 1
　1.2 芒果的营养价值与功能 ·· 2
　1.3 芒果的食用方法与储藏技巧 ·· 3
　1.4 芒果的分类与主栽品种 ·· 5
　1.5 芒果的形态特征与生长环境 ·· 12
　1.6 芒果的挑选与储藏方法 ·· 16

2 全球芒果产业格局与我国芒果生产情况 ·· 18
　2.1 全球芒果产业格局 ·· 18
　2.2 部分芒果主产国生产情况 ··· 20
　2.3 我国芒果产业情况 ·· 24

3 芒果种植管理技术 ·· 28
　3.1 芒果种苗及其繁殖技术 ·· 28
　3.2 芒果栽培管理技术 ·· 36

4 芒果主要病虫害防治与采后保鲜技术 ··· 48
　4.1 芒果主要虫害及其防治 ·· 48
　4.2 芒果主要病害及其防治 ·· 53
　4.3 芒果采收及保鲜技术 ··· 59

5 我国芒果产业现状与发展展望 ································ 66

5.1 我国芒果产业现状 ································ 66
5.2 我国芒果产业发展情况 ································ 66
5.3 我国芒果产业发展存在的问题和风险 ································ 69
5.4 我国芒果市场与产业前景分析 ································ 71
5.5 展望 ································ 72

CONTENTS

1 Introduction ································ 73

1.1 The overview ································ 73
1.2 Nutritional value and function ································ 74
1.3 Edible methods and storage skills ································ 76
1.4 Classification and main varieties of mangoes ································ 78
1.5 Mango trees' morphological characteristics and growing environment ································ 87
1.6 Selection and storage methods ································ 92

2 Global Mango Industry Pattern and Chinese Mango Production ································ 95

2.1 Global mango industry pattern ································ 95
2.2 Production situation of some main mango producing countries ································ 97
2.3 Mango industry in China ································ 103

3 The Planting Management Techniques ································ 108

3.1 Mango seedlings and their propagation techniques ································ 108
3.2 Mango cultivation and management techniques ································ 119

4 Main Pest Control and Post-harvesting Preservation Technology of Mango ································ 136

4.1 Main pests and control of mango ································ 136
4.2 Major diseases of mango and its prevention and control ································ 142
4.3 Mango harvesting and fresh-keeping technologies ································ 151

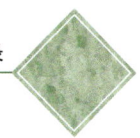

5 Current Situation and Future Development of China's Mango Industry ········· 159

5.1 Current situation of China's mango industry ············· 159
5.2 Development of China's mango industry ·············· 160
5.3 Current problems and risks in China's mango industry ········· 163
5.4 Mango market and industry prospect in China ············ 165
5.5 Outlook ································· 167

参考文献 ····································· 169

1 绪 论

1.1 芒果概述

芒果（*Mangifera india* L.）属漆树科（*Anacardiaceae*）杧果属（*Mangifera*），是杧果的通俗名，被誉为"热带果王"（图1-1）。全球种植面积640.7万 hm^2，产量超过5 600万 t（FAO，2020），仅次于柑橘、蕉类、葡萄、苹果，居世界水果第五位。

图1-1 芒果

芒果原产于印度，现广泛分布于南北纬30°之间，冬季最冷月均温11 ℃以上，绝对低温3.7 ℃以上的热带、亚热带地区，北至我国四川南部和日本南部岛屿，南至非洲南部。全世界有超过100个国家栽培芒果，主要生产国有印度、泰国、中国、印度尼西亚、菲律宾、墨西哥、巴基斯坦、尼日利亚、埃及、科特迪瓦等。主要出口国有墨西哥、巴西、厄瓜多尔、秘鲁、印度、巴基斯坦、泰国、菲律宾和中国，主要进口国家有美国、加拿大、沙特阿拉伯、阿拉伯联合酋长国、科威特、日本、新加坡、英国、法国、俄罗斯、荷兰、比利时、德国等。

芒果具有适应性广、速生易长、抗逆性强、结果早（植后2~3年开始结果）、栽培管理容易、产量较高、经济寿命长（50年以上）等优点，果实形态美观，色泽诱人，果肉质地细滑多汁，香味独特，风味浓郁，营养丰富，极受消费者欢迎。

1.2 芒果的营养价值与功能

1.2.1 芒果的营养价值

芒果肉质细腻，气味香甜，营养价值很高，含有丰富的蛋白质、脂肪、碳水化合物、膳食纤维、钙、铁、磷、钾、钠、铜、镁、锌、硒、锰、维生素 B_1、维生素 B_2、烟酸、维生素 C、维生素 E、维生素 A、胡萝卜素、没食子酸、槲皮素、芒果酮酸、异芒果醇酸、阿波酮酸、阿波醇酸、芒果甙、类黄酮、超氧化物歧化酶、β-隐黄质、番茄红素、丁香酸、槲皮素等（图1-2）。集热带水果精华于一身，食用芒果还有滋润肌肤、降压降脂等作用。

图1-2 芒果果肉

1.2.2 芒果的健康功效

芒果性凉，味甘酸，入肺、脾、胃经。有益胃、解渴、利尿的功用。主治口渴咽干、食欲不振、消化不良、晕眩呕吐、咽痛音哑、咳嗽痰多、气喘等病症，还具有很好的保健功能。

抗癌：芒果含有大量的超氧化物歧化酶及类黄酮等多种活性成分，这些活性成分具有防癌、抗癌的作用。

防治心脑血管疾病：芒果含有维生素、矿物质等，除了具有防癌的功效外，同时也具有防止动脉硬化及高血压的食疗作用。芒果中维生素 C 含量高于一般水果，且具有加热加工处理后维生素 C 也不会消失的特点。常食芒果不仅可以补充体内维生素 C 的消耗，也可降低胆固醇、甘油三酯，有利于防治心脑血管疾病。

祛痰止咳：芒果中所含的芒果苷有祛痰止咳的功效，对咳嗽痰多气喘等症有辅助治疗作用。

护目养颜、延缓衰老：芒果的糖类及维生素含量非常丰富，尤其维生素 A 含量为水果之首，具有明目的功效。

健胃止晕：芒果具有清肠胃的功效，晕车、晕船时食用芒果具有一定的止吐作用。

抗菌消炎：芒果未成熟的果实、树皮、茎能抑制化脓球菌、大肠杆菌等，芒果叶的提取物也同样有抑制化脓球菌、大肠杆菌的作用，可治疗人体皮肤、消化道感染。

1.3 芒果的食用方法与储藏技巧

1.3.1 芒果的食用方法

芒果成熟后直接食用是芒果最普遍的食用方法，能最大限度地保留芒果的口感风味和营养价值。把成熟的芒果洗净，然后用水果刀纵向从头到尾划一刀，紧贴着果核把芒果切成两半。将有果核的一半再切一刀使果核剖离，再把果核上连着的果肉剔掉。接下来在吃之前，用刀在芒果纵剖面上纵向、横向再划几刀，注意不要划破果皮。吃的时候只要用手指向上一顶，就可以方便地吃到可口的芒果了（图 1-3）。

广东和广西等地的人们喜欢把芒果腌酸后食用；有的人喜欢把生的芒果去皮，果肉加入辣椒等调料食用。

芒果可以通过深加工制成芒果干、芒果果脯、芒果酱、芒果罐头等。

图1-3 金煌芒果肉切面

1.3.2 食用芒果的注意事项

芒果好吃,但不宜多食。芒果糖分高,每100 g果肉含糖约12.9 g,摄入过多芒果容易引起高血糖,因此糖尿病患者不宜食用;芒果果肉中钾含量很高,因此不宜大量食用;肾病患者大量食用芒果易加重慢性肾病的症状,女性经期也不宜食用。

芒果香甜润滑的口感深受消费者喜爱,但也有一些人群对芒果过敏。芒果中含有果酸、氨基酸以及某些刺激皮肤的物质。另外,不完全成熟的芒果中还含有醛酸,会对皮肤黏膜产生刺激从而引发过敏。在吃芒果时很容易将芒果汁沾到嘴角、脸颊等部位,会刺激面部皮肤,造成面部红肿、发炎,严重者会出现眼部红肿、疼痛现象。

对于大部分人来说,食用芒果过敏是因为不良习惯造成的。一些小朋友或成人吃芒果时,习惯于剥开皮,直接咬着吃。但芒果果皮中的刺激物较多,如用手剥完后应及时洗手,或戴上一次性手套处理果皮以防止刺激物刺激皮肤。用刀削皮的时候可以多削一点皮下来。也可用勺子将果肉送到口中,使嘴周边及面部尽量不接触芒果汁液。

吃芒果过敏后可以用冰水冲洗皮肤,清洗掉皮肤上残留的芒果汁。一般来说,吃芒果过敏会出现红肿、发痒的症状,也会出现烧灼的感觉,用冰水冲洗可以镇静皮肤,减轻红肿烧灼的感觉,也能起到很好的止痒作用。

如果连续多次吃芒果都出现过敏症状,说明是过敏体质,那么不建议继续食用芒果,也建议避免接触芒果的树干、茎叶等,尤其要避免接触到芒果树的汁液。

1.4 芒果的分类与主栽品种

1.4.1 芒果的分类

全世界的芒果栽培品种有1 000多个,从植物学分有单胚和多胚两大种群,我国商业化栽培的有40余个品种,主要分布于海南、云南、广西、广东、福建、台湾等地。

单胚类型的芒果种子仅有一个胚,播种后仅出一株苗,实生树变异性大,不能保持母本优良性状,如印度芒及其实生后代(如红芒类),中国的紫花芒、桂香芒、串芒、粤西1号和红象牙等均属单胚品种。多胚类型芒果种子有多个胚,播种后能长出几株苗,能发育成苗的胚多属无性胚,故实生树变异性小,多数能保持母本性状,如菲律宾、泰国的许多品种及海南省的土芒多属于这一类型。

1.4.2 几种国内芒果主栽品种

1.4.2.1 金煌芒

由中国台湾黄金煌先生选育,所以命名为金煌芒,在1999年引进至中国大陆地区栽培,目前主要分布在广东、广西、云南、福建、台湾等省(区)。金煌芒的特点是单果重量大,单果重500~1 250 g,最大可达2 400 g。果为长卵形,皮厚,成熟时为金黄色,核小、纤维少、汁多,风味清甜,可食率80.1%,可溶性固形物17%~19%,含糖度值17%,品质优,商品性好,耐储运(图1-4)。

品种特性:树势强,树冠高大,花朵大而稀疏,产量极高,一般亩[①]产可达2 000~3 000 kg,对炭疽病具有一定的抗性。

不足之处:易发生"水泡病"和一些生理病害。

图1-4 金煌芒

---

① 1亩≈667m², 15亩=1 hm²。

1.4.2.2 贵妃芒

品种来源：又名红金龙、金凤，由我国台湾地区选育而来，目前已成海南的主栽品种之一，在云南、广东、四川也有广泛种植。

品种特性：该品种树势健旺，树冠呈圆头形，叶片宽且长，早产丰产，4~5 年生嫁接树单株产量为 20~30 kg 或者更高，果实长椭圆形，果顶较尖小，果实有大小之分，小的贵妃芒单果重约 150 g，大的重 500~600 g（图 1-5）。

优势：果品光滑呈绯红色，成熟后红黄色，无任何斑点，艳丽吸引人，果肉橙黄，核小无纤维，水分充足，可食率 74.5%~84.3%，可溶性固形物 14%~17%，甜度可达 14~18 度，品质优。在收获期天旱而光照充足时，果实较耐储运。该品种知名度高，在市场颇受消费者的欢迎。

图 1-5　贵妃芒

1.4.2.3 台农一号

由我国台湾地区凤山园艺所选育，是目前国内栽培面积最大的一个品种。

品种特性：树冠粗壮，生势壮旺，直立，开花早，花期长，抗风抗病性强，两性花比例较高，丰产稳产，较抗炭疽病，适应性广。一般嫁接苗定植后 3 年单株产量可达到 5~10 kg 或更高。

优势：单果重 150~200 g，果品青绿色，果肩向阳面带胭脂红色，成熟时果色金黄，外观美丽，果肉深黄色，组织较细密，味甜，纤维少，质地较细滑，品质好，华南热带农业大学研究结果表明：台农一号的可溶性固形物16.8%，总糖16.76%，可食部分60.6%，耐储运，货架寿命长（图1-6）。

不足之处：台农一号的缺点是果偏小，果实套袋时人工及生产成本较其他品种偏高。

1.4.2.4 肯特芒

原产美国佛罗里达州，由中国热带农业科学院南亚热带作物研究所于1984年引入。果实卵圆形，长约10.8 cm，宽约10.0 cm，厚约9.2 cm，平均果重447 g；成熟果皮黄绿至杏黄色，盖色深红；果肉橙黄色，质地细滑，多汁，纤维少，风味浓郁且香甜；种子小，单胚；果实成熟期7—8月。较耐储运。该品种产量较高，枝条密集，但抗风性较差，适宜在无台风地区种植。现为四川、云南金沙江干热河谷主栽品种（图1-7）。

图1-6　台农一号

图1-7　肯特芒

1.4.2.5 凯特芒

原产美国佛罗里达州，为晚熟品种，由中国热带农业科学院南亚热带作物研究所于1984年引入，目前在我国台湾、广西、云南及四川攀枝花等地大量种植。

品种特性：树势中等，枝条开张，花期较迟，以高产、优质、丰产、稳产、迟熟著称，叶片浓绿，抗风抗寒力强，果实呈卵圆形，有明显的果鼻，果实个头大小不一，平均单果重680 g，大者可达2 000 g以上，未成熟的果实灰紫绿色，成熟后暗红色，有淡紫色的果粉。该品种特别适合于夏季无台风地区及高温干旱地区种植，现为四川攀枝花地区的主栽品种之一（图1-8）。

优势：皮薄、核小、肉厚，纤维少，果肉橙黄，含糖量可达14%~19%，可食率75%左右，接近果核的部位略带酸味。凯特芒为晚熟品种，一般台农一号、贵妃、金煌、椰香等早熟品种完全下市后才开始上市，刚好填补这一时期芒果市场的空档，市场销量很好。

不足之处：在部分地区产量不够稳定，受其他晚熟品种市场竞争的影响，近年来种植面积在逐渐下降。

图1-8 凯特芒

1.4.2.6 爱文芒

也译称爱尔文芒、欧文芒，产自美国佛罗里达州，为第三代红芒，1954年引入我国台湾后又名苹果芒，成为台湾的主要栽培品种，1984年自澳大利亚引入中国热带农业科学院南亚热带作物研究所。在攀枝花红格乡、海南昌江大量种植。

品种特性：树势比较矮小，适于密植，树冠圆头型，花序圆锥形，花梗浅红色，花穗大，坐果率高，花期在1—2月，一般可以出3批花，丰产稳产性好。

优势：果实倒卵形，青果紫绿色或蟹青色，盖色紫红色；成熟的果实底色深黄，盖色鲜红，色泽鲜艳、被誉为芒果中的红宝石。果肉黄色，肉质腻滑，纤维少，味甜，品质较好，成熟时呈金黄色，无纤维，含糖14%~16%，可溶性固形物15%~24%，味道香甜，口感极佳，深受消费者的欢迎（图1-9）。

不足之处：产量受气候的影响较大。

图1-9　爱文芒

1.4.2.7　海顿芒

原产美国佛罗里达州，果实卵圆形，平均果长10.4 cm，宽9.6 cm，厚8.9 cm，单果约重350 g，成熟时果皮橙黄色带鲜红晕，果肉橙黄色，果肉质地稍粗，纤维少，味甜，香味浓，品质上等，果成熟期6—7月，种子单胚，该品种宜在高温干旱且阳光充足的地方栽培，在多雨地区栽培产量和颜色不稳定，果实耐储运，该品种也为南非、墨西哥、以色列等国的主栽品种（图1-10）。

图1-10 海顿芒

1.4.2.8 白象牙芒

原产泰国,是该国主要出口品种之一,现为海南和云南的主要商业栽培品种之一。

品种特性:该品种树势强壮。桂冠高圆头形,干枝分枝较小,直立性强。开花期3月下旬至4月下旬,花序圆锥形,花序轴淡红色。果实硕大,长卵圆形,果弯明显,果嘴痕迹明显。果形似初生象牙,故名象牙芒(图1-11)。

优势:单果平均重量680 g,大者可达2 000 g以上,果实成熟时呈金黄色,皮薄、核小、肉厚、纤维少、果肉橙黄,糖分含量17%,鲜嫩、多汁,味美可口,香甜如蜜。果实较耐储运。

图1-11 白象牙芒

1.4.2.9 南多美芒

又称青皮芒，泰国白花芒，原产泰国，我国云南和海南栽培较多。

品种特性：树势中等强壮，叶革质，互生；花小，黄色或淡红色，成顶生的圆锥花序。

优势：果肾形，有明显的腹沟，成熟果皮暗绿色至黄绿色。果肉淡黄色至奶黄色，肉质细腻，皮薄多汁，有蜜味清香，纤维极少，种子扁薄，多胚。单果重200~300 g，可食部分占72%左右，品质极优，为理想的鲜食品种（图1-12）。

不足之处：该品种花期遇低温阴雨会出现花而不实，并且易裂果，产量中等，植株易感流胶病，其果皮青色，肉色淡，在一些地方影响其销路和价值。

图1-12　南多美芒

1.4.2.10 紫花芒

由广西大学农学院从泰国芒的实生后代中选育，在广西和广东广泛栽种。

品种特性：树势中等，枝条开展，早结、丰产、稳产性较好。嫁接苗植后3~4年结果，6龄树亩产可达1 000 kg或更高。花序圆锥形，花序轴紫红色。花期晚，3月下旬至4月开花，能避过低温阴雨期，较耐修剪，适于矮化密植栽培。

优势：果实斜长椭圆形，两端尖，果皮灰绿色，向阳面淡红黄色，经后熟后转为鲜黄色，果实表皮蜡粉较厚；单果重250~300 g，外形美观，果肉黄色，肉质较细滑，甜酸适中，芳香，香味稍淡，纤维极少或无，果汁多，可食率64%~73%，可溶性固形物13%~15%，糖含量12~15 g/100 mL，酸含量0.09~0.65 g/100 mL，较耐储运（图1-13）。

图1-13　紫花芒

不足之处：该品种对低温反应敏感，不抗寒，易受寒害。

1.4.2.11　R2E2

又称澳芒，原产澳大利亚，由中国热带农业科学院南亚热带作物研究所于1997年引进，平均单果重716 g，品质优良，为澳大利亚主栽品种，在我国台湾、海南、广西、云南有大量种植。

品种特性：该品种粗生易管，早结丰产。

优势：澳芒单果重量为500~1 500 g，形状类似苹果，外表光滑靓丽，颜色呈现金黄色带有红色霞晕，香味非常浓郁，肉质鲜美，清甜可口，无纤维感（图1–14）。

图1–14　R2E2

1.5　芒果的形态特征与生长环境

1.5.1　芒果的形态特征

芒果属常绿大乔木，自然生长条件下一般高10~20 m，树皮灰褐色，小枝褐色，无毛。叶薄革质，常集生枝顶，叶形和大小变化较大，通常为长圆形或长圆状披针形，长12~30 cm，宽3.5~6.5 cm，先端渐尖、长渐尖或急尖，基部楔形或近圆形，边缘皱波状，无毛，叶面略具光泽，侧脉20~25对，斜升，两面突起，网脉不显，叶柄长2~6 cm，上面具槽，基部膨大。

圆锥花序，长20~35 cm，多花密集，被灰黄色微柔毛，分枝开展，最基部分枝长6~15 cm，苞片披针形，长约1.5 mm，被微柔毛，花小，杂性，黄色或淡黄色，花梗长1.5~3 mm，花瓣长圆形或长圆状披针形，长3.5~4 mm，宽约1.5 mm，无毛，里面具3~5条棕褐色突起的脉纹，开花时外卷，花盘膨大，雄蕊仅1个发育，长约2.5 mm，花药卵圆形，子房斜卵形，直径约1.5 mm，无毛，花柱近顶生，长约2.5 mm。

芒果果实形态有椭圆形、肾脏形及倒卵形等。成熟后果皮呈绿色、黄色或紫红色，果肉为黄色或橙黄色，果汁及纤维含量因品种而异。

核果大，肾形（不同栽培品种形状差异大），压扁，长5~10 cm，宽3~4.5 cm，成熟时黄色，中果皮肉质，肥厚，鲜黄色，味甜，果核坚硬。

1.5.2 芒果的生长习性

枝梢生长习性。芒果枝梢呈蓬次式生长，芽由苞片包裹，生长时苞片先绽开，芽梢伸长，叶片开展，苞片随即脱落。中、下部叶片互生，叶距较大。一般苗期和幼树每年抽6~8次梢，幼龄结果树抽2~4次，成龄树1~2次。3—5月抽生的枝梢为春梢，6—8月为夏梢，9—11月为秋梢，12月至翌年2月为冬梢。在海南秋梢是主要结果母枝，但春梢、夏梢也可成为结果母枝，在条件良好的情况下，某些品种在12月至翌年1月抽生的冬梢也能开花结果。从芽萌动至枝梢停止生长、叶片老熟历时15~35 d。夏梢、秋梢历时较短，冬梢较长。枝梢生长与根系生长交替进行（图1-15）。

叶芽

嫩叶　　　　　　　　　老熟叶片

图 1-15　芒果抽生的叶芽、嫩叶与老熟叶片

花芽分化。正常情况下，芒果花芽分化从 10 月下旬至 11 月开始。用药剂催花则可以使其花芽分化不受季节限制。从花芽分化至花序的第一朵花开放历时 20~39 d，但第一朵花开放后花序仍可继续伸长。适当的低温干旱有利于花芽分化，气温高有利于两性花的形成。

开花。芒果树自然开花在每年 12 月至翌年 2 月，有时会早至 11 月或迟至翌年 3 月，盛花期在春节前后。单个花序从第一朵花开放至全花序开放需 15~25 d，一株树的花期约 50 d（图 1-16）。芒果花分两性花与雄花，两性花有发育正常的雄蕊和雌蕊，可进行正常的传粉受精和结实，雄花没有雌蕊，开花后不能结实。多数栽培品种两性花占 15% 以上。一朵花由花瓣展开至柱头干枯约 1.5 d。

图 1-16　芒果花

果实。开花受精后子房开始膨大，约 1.5 个月后迅速增大，采果前 10~15 d 果实个头增长极缓慢或不增长，此时主要是果实增厚、充实及重量的增加。从开花稳实至果实青熟，早熟品种需 85~110 d，中熟品种需 100~120 d，晚熟品种需 120~150 d（图 1-17）。在果实发育期间有两次明显的落果高峰：第一次在花后两周左右，主要是受精不良的小果枯黄脱落，落果量较大；第二次在花后 4~7 周，除小部分是发育不良的畸形果或败育果外，更多为养分和水分不足造成落果。花后 2.5 个月后很少再发生生理性落果，我国常规栽培的品种果实收获期在 5—7 月，具体因品种和地区而异。

图1-17 芒果开花后坐果

1.5.3 芒果生长的环境要求

温度。芒果性喜温暖,不耐寒霜,最适生长温度为25~30 ℃,低于20 ℃生长缓慢,低于10 ℃叶片、花序会停止生长,近成熟的果实会受寒害。芒果生产区年均温在20 ℃以上,最低月均温大于15 ℃。温度不足会造成授粉受精不良,甚至花序枯死或种胚败育死亡。气温高于37 ℃时,小花和果实产生日灼,低于10 ℃,新梢及花穗停止生长,5 ℃以下,幼苗、嫩梢和花穗受寒,0 ℃左右,幼苗地上部、成年树的花穗和嫩梢、外围叶片都会受害,严重时枯死。-3 ℃以下幼树冻死,大树严重受冻。

光照。芒果为喜光果树,充足的光照可促进花芽分化、提高开花坐果率和果实品质。通常树冠的阳面或空旷环境下的单株开花多,坐果率高,枝叶过多、树冠郁闭、光照不足的芒果树开花结果少,果实外观和品质差。可通过整形修剪,改善园内、树内的透光条件以提高产量和延长盛产期(图1-18)。

水分。芒果在年降水量700~2 000 mm的地区生长良好。花期和结果初期如空气过分干燥,易引起落花落果,雨水过多又易导致烂花和授粉受精不良,夏季降雨过于集中,常诱发严重的果实病害,采收后的秋旱会影响秋梢母枝的萌发生长。

图1-18 芒果果园

土壤。芒果对生长土壤要求不高，但以土层深厚，地下水位低于3 m以下，排水良好，微酸性的壤土或沙壤土为好。

1.6 芒果的挑选与储藏方法

1.6.1 如何挑选芒果

看软硬。按压表皮可感受果肉有下陷变软则为成熟，较硬则未熟。

看颜色。大部分品种芒果成熟后皮色为橙黄色，未熟时是青绿色，切开芒果，果核硬，果肉呈黄色，为熟果，若果核软，果肉为白色或绿色则未熟（图1-19）。

看比重。将芒果放入水中测试，沉下去的说明熟了，浮在表面的则未成熟。

图1-19 贵妃芒果实

闻味道，若带有酸甜果香味则为成熟芒果，而带有青涩味或无味则未成熟。

1.6.2 如何区分树熟与催熟芒果

看颜色。自然成熟的芒果颜色不均匀，而催熟的芒果只有小头尖处果皮翠绿，其他部位果皮均发黄。

闻香味。自然成熟的芒果大多能闻到一种果香味，催熟的芒果闻着味淡或有异味。

看软硬。自然成熟的芒果弹性较好，但催熟的芒果整体较软。

看食用感受。催熟的芒果在吃的时候水分不多、糖分不够，如果用手直接撕开果皮，会感觉到果皮紧黏着果肉，很难成块撕开。树上熟的芒果汁水丰富，甜度较高，果皮容易与果肉分开，成熟度高的芒果致敏成分也较少，食用后不容易发生过敏现象及其他身体不适症状。

看粉层。树熟芒果在果皮表面有一层灰白色的粉层，催熟的芒果没有此粉层。

1.6.3 如何储藏芒果

芒果属于呼吸跃变型水果，果实适宜六至八成熟时采收，此时芒果质硬、味涩，需经过后熟（即果肉所含的淀粉转化为可溶性糖类）才可以食用。芒果常温下储藏期一般只有 7~12 d，适度低温储藏可以延长芒果的保鲜期，市面上常用 10~13 ℃的冷库储藏，低于 10 ℃时，芒果容易遭受冷害而无法后熟。因此，芒果还没熟透不适宜放在冰箱中过低温度冷藏，否则无法后熟变软。当芒果变软、熟透后，就可以放冰箱保存了。但是保鲜期不会太长，因为即使冰箱冷藏也无法阻止采后病害的发展。炭疽病和蒂腐病是芒果采后最主要的两大病害。芒果储藏中后期，果皮上出现的黑色斑块就是采后病害发展的表现，病斑会随着时间延长而不断扩大，甚至果肉腐烂渗出黑褐色汁水。

避光、阴凉、通风干燥的环境下更利于芒果的保存。采果时留 2~3 cm 的果梗，芒果的枝蒂处要尽量保持干燥完整，避免果胶粘到果实上。如果想让芒果尽快成熟变软，可将芒果与成熟香蕉放在一起，利用香蕉释放的乙烯气体更快催熟芒果。

芒果熟透后可以制成果酱，或者将外皮削掉后整个放入冰箱冰冻，此法可长期保存芒果，解冻后食用，别有一番滋味。

制干保存。对于吃不完的熟芒果，可以采用冻干或晒干的方式，脱去芒果多余的水分，然后密封干燥保存，芒果干一般可存放 6 个月以上。

2 全球芒果产业格局与我国芒果生产情况

2.1 全球芒果产业格局

芒果是世界五大水果之一,其产量仅次于葡萄、柑橘、香蕉、苹果,生产规模在热带水果中排名第三位。据联合国粮食及农业组织(FAO)统计,现有103个国家生产芒果,主要集中在亚洲、南美洲和非洲,涵盖了北起我国四川南部,南至南美洲,横跨南北纬30°之间的地区。亚洲是芒果种植面积最大的地区,亚洲芒果总产量约占世界芒果产量的85%;其次是美洲,产量约占世界总产量14%。世界主要芒果生产国的芒果成熟季节见表2-1。2020年全球芒果生产排名前十的国家如下。

印度:印度是世界芒果收获面积最大的国家,收获面积近200万hm^2,每年生产芒果约1 633.74万t,占世界芒果总产量的42.2%,印度的芒果种类繁多,被认为是芒果种类最多的国家。

中国:我国是全球第二大芒果生产国,截至2020年,全国芒果种植面积达34.94万hm^2,总产量330.6万t,产量约占全球的总产量的8.75%,产值达205.2亿元。

泰国:每年生产约255.05万t芒果,占世界芒果总产量的6.5%,排在全球芒果生产的第三位。

巴基斯坦:每年芒果产量约178.43万t,占世界芒果总产量的4.6%,排在全球芒果生产的第四位。

墨西哥:每年生产约163.27万t芒果,占世界芒果总产量的4.2%,排名全球芒果生产的第五位。

印度尼西亚：每年芒果产量约为131.35万t，占世界芒果总产量的4.1%，印度尼西亚主要芒果产区包括东爪哇、南苏拉威西岛、东加里曼丹、西努沙登加拉省，是世界上第六大芒果生产国。

巴西：世界上第七大芒果生产国，芒果年总产量为118.89万t，占世界芒果总产量的4%。

孟加拉国：世界上第八大芒果生产国，每年生产约104.79万t芒果，占世界芒果总产量的3.9%，由于其气候适宜，孟加拉国市场几乎全年都提供新鲜多汁的芒果。

菲律宾：世界芒果生产国中排名第九，每年生产约82.36万t芒果，占世界芒果总产量的3.6%。

尼日利亚：在世界芒果生产国中排名第十，每年生产约79.02万t的芒果，产量占世界芒果总产量的3%。

表2-1 世界主要芒果生产国的芒果成熟季节

生产区域		上市时间
中国	海南	3—5月（11月至翌年3月海南南部有早果，6月有个别晚果）
	广东	6月中旬至8月上旬
	广西	5—9月（其中6—7月集中采摘）
	云南	5—8月（9—11月有少量）
	攀枝花	8—11月（6—7月已有早果）
以色列		8—9月
巴西		9—12月
秘鲁		1—2月
澳大利亚		12月至翌年2月
南非		1—3月
菲律宾、泰国		1—5月
越南		9月至翌年6月（高峰期在2—3月）
马里、科特迪瓦、布基纳法索		3—5月
印度、墨西哥、委内瑞拉、波多黎各、哥斯达黎加		5—7月（高峰期）

2.2 部分芒果主产国生产情况

2.2.1 印度

芒果在印度有 4 000 年以上的栽培历史,是公认有记载最早种植芒果的国家,也是目前世界上芒果栽培面积最大、总产量最高的国家。2016 年,印度芒果栽培面积约为 212 万 hm^2,产量 1 900 万 t,单产 8.9 t/ hm^2。印度共有 14 个邦生产芒果,栽培面积从大到小排列为:安德拉邦、北方邦、奥里萨邦、卡纳塔克邦、特伦甘纳邦、泰米尔纳德邦、马哈拉斯特拉邦、古吉拉特邦、比哈尔邦、西孟加拉邦、恰蒂斯加尔邦、喀拉拉邦、恰尔康得邦、中央邦。单产最高的为北方邦,达到 17.14 t/ hm^2;单产最低的为马哈拉斯特拉邦,仅为 3.28 t/ hm^2。印度芒果种植的品种多达数百个,收获期在 3—8 月,主要的商业品种包括阿方索(Alphanso)、Amrapali、Aswina、Banganpalli、BombayGreen、Chausa、Chenlkurasam、Chinnarasam、Dashehari、Fazli、Gulabkhas、Himayuddin、Himsagar、Kalpadi、凯撒(Kesar)、Kishen Bhog、LakshmanBhog、Langra、Mallika、Mulgoa、Mundappa、秋芒(Neelum)、Pairi、Poiri、Rajapuri、Ramkela、RaniPasanddegn Rataul、Rumani、Sepia、Sukul、Sunderij、suvarnarekha、Totapuri、vanraj 和 zardalu,其中阿方索芒果种植面积最大,产量最高。

在育种方面,近年来印度从杂交后代材料中培育出 Mahmood Bahar 等 28 个品种,实生选育出 Pusa surya 等 7 个品种。印度的芒果果苗繁殖主要采用嫁接法,分为枝接和靠接。种植采用高密度种植技术,包括矮化砧木、树冠控制、施肥和多效唑的使用而获得早期的高产。如 Amrapali 种植密度为 1 600 株 /hm^2,Dashehari 种植密度为 3 m×2.5 m。

在树体管理方面,印度主要通过修剪技术及使用多效唑控制树高和树势,部分品种修剪后依然保持丰产稳产,部分品种大小年现象明显。如 Banganpalli、Suvarnarekha、Neelum 和 Bangalora 等品种对修剪十分敏感,通过周年或隔年修剪可控制树冠,开张型树冠,有利于坐果和提高果品质量。多效唑目前在印度被广泛使用来控制树势和促进成花,通常通过土壤淋施和叶面喷施来进行,土壤施用浓度为每棵树 2.5~10 g,叶面喷施浓度为 0.5~2.0 g/L,主要根据树龄和树势而定。土壤淋施效果更好,施用时间在 7—10 月。在保果方面,主要使用萘乙酸和赤霉素。Mallika 推荐萘乙酸浓度为 20 mg/kg,Dashehari 推荐赤霉素浓度为 10 mg/kg。在间作方面,早期可种植花生、生姜、菠萝及马铃薯等。在保鲜方面,目前主要采用低温储藏(温度在 7~12 ℃,相对湿度在 85%~95%)和气调储藏。在果实蝇控制方面,主要利用果蝇诱捕器诱捕果实蝇。

2.2.2 泰国

泰国位于中南半岛，享有"水果王国"的美誉。泰国是世界上的主要芒果生产国和出口国，气候条件非常适合芒果生产，芒果品种繁多。泰国果农的生产栽培技术水平较高，芒果产期调节已成为泰国芒果生产上普遍应用的技术。由于种植芒果收益较高，因此在所有经济果树种类中芒果的栽培面积最大，果农数量较大。泰国 96% 的芒果用于鲜食，1.17% 用于加工，仅 2.23% 用于出口。2017 年泰国芒果出口数量约 5.6 万 t，主要鲜果出口市场为亚洲的韩国、日本、中国及欧洲国家。

泰国各地均有商业化芒果生产。主栽品种 Namdokmai 主要种植于泰国北部（彭世洛府、清迈府）、东北部（呵叻府、黎府）、东部（差春骚府、沙橄府）及南部（巴蜀府）。

自然情况下，泰国主要的芒果收获季节是 4—5 月。泰国芒果的生长发育季节刚好是泰国的旱季，因此果实品质很好。泰国芒果成熟时间依据纬度不同，从中部地区最先开始成熟，然后依次是东部、北部、东北部和北部高山地区。此外，泰国近年来发展的反季节生产技术能使芒果实现周年收获。

泰国有芒果品种超过 200 个，商业化栽培品种大约有 10 个。根据食用习惯，泰国的芒果可分为三大类。第一类为生食芒果，这类芒果在未成熟阶段就可以收获和食用，它们有独特的生脆口感，不酸或稍带酸味。生食芒果的品种有 Khieu Savoi、Raet、Falon 等。第二类为熟食芒果，这类芒果在完全成熟时采收，后熟软化后食用，品种有 Namdokmai、Namdokmai Si Thong、Mahachanok、Okrong 等。第三类为加工型芒果，用于加工成罐头、芒果干、芒果卷、腌芒果、芒果汁、芒果冰激凌等，较出名的品种有 Kaeo、Mahachanok、Chokanan 等。近年来，泰国也逐渐从澳大利亚、中国台湾等国家和地区引进 R2E2、玉文等果皮颜色鲜艳的芒果品种进行试种栽培，其中 R2E2 由于果实外观、果实风味及产量较好，栽培面积也逐渐扩大。

泰国通常采用种植嫁接苗的方式进行芒果生产，商业化的种苗生产主要采用靠接方式进行嫁接。靠接使穗和砧木同时保留各自的根系，因此嫁接的成活率较高。另一种方式是先在果园种植 1~1.5 年的实生苗，然后再进行嫁接，这种方式比购买嫁接苗的种植成本更低。

泰国芒果常规栽培的株行距一般为 10 m×6 m，种植坑规格一般为 30 cm×30 cm×30 cm。而在中部地下水位高的地区，果农通常采用深沟高畦方式种植，一般畦宽 6~8 m，沟宽 1~1.5 m。一般采用 2.5 m×2.5 m 的密植栽培方式。这种矮化密植方式有利于修剪、打药、疏果、套袋、收获等作业的进行，单位面积的产量高于常规栽培模式。

植株整形修剪通常是在采果后进行，修剪回缩结果枝 60 cm 左右，还将老弱枝剪除以使树体通风透光良好。农民通常会将修剪后的枝条放在根部作为覆盖物。在芒果树开花挂果期间，当花序长到 2~3 cm 长时，农民会使用杀虫剂控制害虫，在盛花期则停止使用杀虫剂以利于昆虫进行授粉。在芒果开始坐果直至果实套袋阶段又恢复使用杀虫剂。在芒果坐果 30~45 d 后，果实大小长约 5 cm、宽约 2 cm 时进行疏果，将太小、畸形或遭受病虫害破坏及败育的果实去除。一般一个花序保留 1~3 个果实以保证果实大小适中。当果实生长到如鸡蛋大小时，果农便进行套袋。套袋除了有防果实蝇的效果外，还能增加果皮颜色，最常用的袋子是里面有黑色涂层的黄色牛皮纸袋。为节约成本，果农通常回收纸袋重复使用。

在泰国，芒果反季节生产有两种方式。一种是利用一年多次开花品种进行生产，另一种是通过化学催花方式。目前利用多效唑来进行芒果反季节催花已成为泰国普遍使用的技术。多效唑通常使用土施方式，施用前先把树盘的杂草清除干净，并保持土壤湿润以提高药剂吸收效率。果农会根据树体健康状态调整多效唑的使用量，通常使用量为树冠直径 10 g/m。使用过多效唑的树在下一年要减少剂量，因为植株可以从土壤中吸收上年残留的药剂。药剂在土壤中的残留量与土壤质地有关，通常排水良好的壤土残留量低于黏土。

泰国果农通常给芒果树施用有机肥和化肥。有机肥通常每年在采果后施用一次，施肥量根据树体大小确定。化肥通常每年施用两次，一次是在采果后和有机肥一起施用，另一次在幼果期作为追肥施用。肥料的施用量由树龄和挂果量确定。此外，果农还使用海藻提取物、钙和硼等叶面肥。芒果树生长需要大量的水分，尤其是在开花和果实发育阶段，在这个阶段通常是泰国的旱季，因此果农需要对芒果树进行灌溉。部分果农通过进行树盘覆盖来保持土壤湿润。

泰国主要的芒果虫害有小黄蓟马（*Scirtothrips dorsalis* Hood）、芒果叶蝉（*Idioscopus clypealis* Lethierry）、绿鳞象甲（*Hypomeces squamosus*）、芒果切叶象甲（*Deporaus marginatus*）、东方果实蝇（*Bactrocera dorsalis* Hendel）等。病害主要有炭疽病（*Colletotrichum gloeosporioides* Penz）、白粉病（*Oidium mangiferae* Berth）、煤烟病（*Meliola mangiferae* Earle）等。泰国部分果农使用病虫害监测技术以控制病虫害发生，如利用黏板监测害虫种群密度，然后确定防治的时机和用药剂量，也使用其他物理防治方法如果实套袋等。为减少农药使用量，果农还使用性诱剂防治果实蝇，以减少果实蝇的种群数量。

4—6 月是泰国芒果自然收获的季节。随着反季节催花技术的普遍应用，目前泰国全年都可以生产芒果。Namdokmai 的果实通常在开花后 105~115 d 或套袋后 55~60 d 时采收。一些果农也通过大小、外形、果皮颜色等来判断芒果的采收时机。

果实供出口的芒果园，通常在果实成熟度为 85% 时采收。采收时，工人小心地摘下果实，避免碰伤。先暂时保留较长的果柄和果袋，采摘完后立即转运到包装棚。在包装棚里，先将果柄剪短至 3~5 cm 长，避免果柄基部果浆流出污染果面，然后去掉果袋。接下来将果实进行分级，挑出符合出口质量要求的果实，余下的将进行再次分级用于国内销售。

满足出口质量要求的芒果，还要进行额外的处理以满足出口要求。先清洗果皮，去掉污渍和灰尘，肉眼检查确保没有炭疽病、害虫、瑕疵及擦伤，再完全去除果柄。接下来进行蒸汽热处理，然后在阴凉处进行吹干，最后进行包装。在果实进行装箱转运到冷藏车里之前还将进行最后的质量检查。

2.2.3 巴基斯坦

芒果为巴基斯坦第二大果树作物，主要出口到中东、英国、阿富汗、东南亚等。巴基斯坦的芒果收获时间为 5—10 月，集中期在 6—7 月。主要种植在旁遮普省和信德省，这两省的芒果产量占巴基斯坦芒果产量的 90% 以上。巴基斯坦种植的芒果品种有 100 多个，主要商业品种为 Anwar Ratole、Baganapalli、Dashehari、Fairi、Gulab Khas、Kala Chaunsa、Langm、Malda、Sindhri、Siroli、Summar Bahist Chaunsa、Suvarnarekha、White Chaunsa 等。其中，Summar Bahist Chaunsa 种植面积最大。

巴基斯坦主要用当地实生单胚品种作砧木，栽培管理技术与印度类似。巴基斯坦芒果害虫主要有芒果蝉、夜蛾、果实蝇、粉蚧、象鼻虫、介壳虫和芒果瘿蚊等，最重要的害虫是芒果果实蝇。巴基斯坦芒果病害主要有黑斑病、炭疽病和白粉病等。巴基斯坦芒果采后损失严重，芒果质量和价格较低。采后损失主要是由于采后处理不合理、合同商利益驱动而进行恶性早采以及运输和储藏设施不足造成的。由于芒果较容易腐烂，芒果商没有及时销售或加工成产品，造成鲜芒果积压受损。巴基斯坦 75% 的芒果商都采用传统的包装方法和当地的化学药物来催熟，导致芒果损伤。针对芒果出口市场，巴基斯坦在芒果收获后采取了以下措施：①通过洗果/打蜡、热水浸泡、冷处理等采后处理措施来提高芒果的货架寿命；②建立芒果出口加工区，以集装箱方式出口；③进行出口前预检；④派销售人员到新兴市场开拓。

2.2.4 越南

芒果是越南种植的主要热带水果之一，仅次于香蕉。2020 年越南芒果种植总面积达 8.7 万 hm^2，总产量达 89.32 万 t，同比增长 6.5%。其中，九龙江三角洲占 48%。

据越南农业与农村发展部农产加工与市场发展局的消息，2020 年，越南芒果出口额超 1.8 亿美元，占世界芒果出口总额的 1.15%。越南芒果主要出口市场是中国（占 83.9%，出口额达 1.52 亿美元），其次为俄罗斯、美国、韩国、欧盟、澳大利亚和日本等。农业与农村发展部设定到 2030 年全国芒果种植面积达 14 万 hm², 产量达 150 万 t, 出口额达 6.5 亿美元, 70% 保鲜加工厂技术达到先进技术等目标。

2.3 我国芒果产业情况

2.3.1 我国芒果产业分布

芒果性喜温暖, 不耐寒霜, 最适生长温度为 25~30 ℃, 低于 20 ℃生长缓慢, 低于 10 ℃叶片、花序会停止生长, 近成熟的果实会受寒害。

我国是世界第二大芒果生产国, 品种资源丰富, 种植区域广泛。我国芒果主要产自海南、广东、广西、云南、四川、福建、贵州等地。其中海南省上市最早, 一般每年的 3—6 月成熟, 然后依次是广东 5—8 月、广西 6—9 月、云南 5—11 月、四川和贵州 7—10 月、福建 8—10 月。

截至 2020 年, 我国芒果种植面积达 34.94 万 hm²（524.1 万亩）, 总产量 330.6 万 t, 产值达 205.2 亿元。近 10 年间, 我国芒果产量保持高速发展, 2011 年我国芒果产量仅为 100.34 万 t, 2020 年我国芒果产量高达 330.6 万 t, 是 2011 年产量的 3 倍以上, 增量达 230.26 万 t, 增幅 229.48%, 年均复合增长率约为 14.17%（图 2-1）。

目前, 我国的热带和亚热带地区均有芒果种植, 主产区分布于海南、广东、广西、云南、贵州、福建、四川和台湾等省（区）的 100 多个市县。台农一号、金煌、贵妃、凯特、桂热芒 82 号、桂热芒 10 号等品种适应性广、品质优, 已成为我国芒果产业的主栽品种。但是, 目前我国自育品种仅占全国芒果种植面积的 15% 左右, 导致同质性严重, 差异化不大。引进品种多、自育品种少也成为制约我国芒果产业发展的深层次原因。

2.3.2 我国芒果产业特点

国内芒果产业可分为 5 个优势产业带, 分别为海南早熟芒果优势产业带、广东雷州半岛早中熟芒果优势产业带、广西右江河谷中熟芒果优势产业带、云南西南—云南南—云南中元江流域芒果优势产业带和金沙江干热河谷流域晚熟芒果优势产业带。不同产区发展的优势品种也有差异。

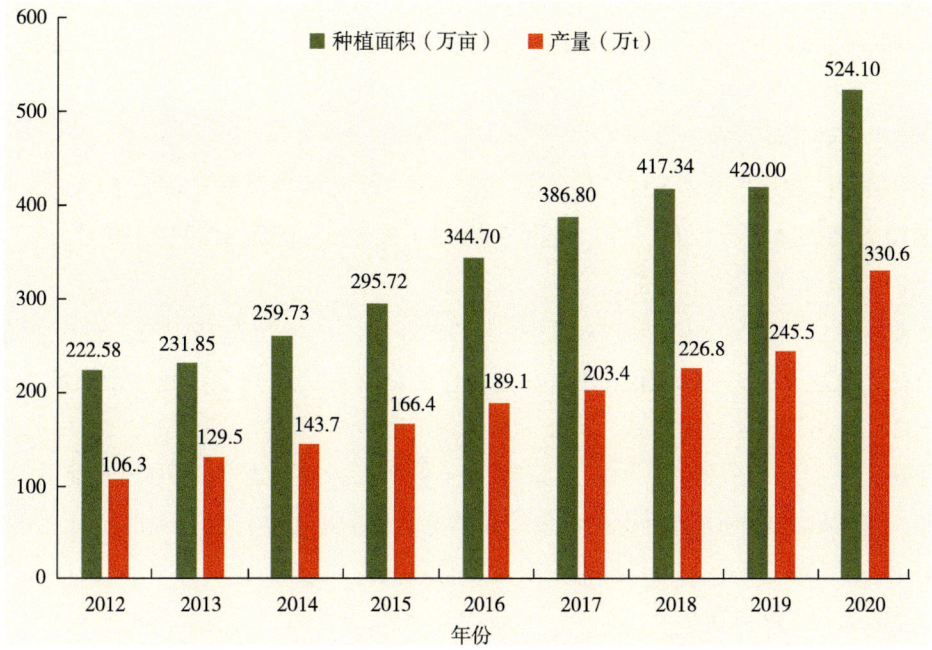

图 2-1　2012—2020 年我国芒果种植面积与产量情况

2.3.2.1　海南早熟芒果优势产业带

种植面积约 87.48 万亩。

品种结构：适宜所有芒果品种生长。主流品种有台农一号、金煌、贵妃、红玉、R2E2 等。

上市时间：11 月至翌年 5 月底，集中上市时期在每年的 3—5 月。

分布概况：种植面积最大的是三亚市，2020 年达到了 36.54 万亩；其次是东方市 21.67 万亩；乐东县排在第三名，为 17.04 万亩；陵水县和昌江县种植面积分别达 6 万亩和 3.78 万亩。

区域品牌："三亚芒果"，中国国家地理标志产品，2018 年入选中国特色农产品优势区名单；2019 年入选中国农业品牌目录；2020 年入选中欧地理标志第二批保护名单。

2.3.2.2　广东雷州半岛早、中熟芒果优势产业带

种植面积约 30 万亩。

品种结构：台农一号、金煌、贵妃、红芒等。

上市时间：6 月中旬至 8 月上旬。

分布概况：最适宜芒果生长区域包括雷州半岛南部的徐闻、雷州、电白、吴川等地。其中，雷州"覃斗芒果"的种植地主要分布在雷州覃斗、乌石、北和、英利 4 个乡镇，种

植面积达 5.2 万亩，年产值约 4.9 亿元。

区域品牌："覃斗芒果"，中国地理标志保护产品。

2.3.2.3 广西右江河谷中熟芒果优势产业带

种植面积约 154 万亩。

品种结构：栽培芒果品种有 30 余个，早、中、晚熟品种面积比例为 5∶4∶1，种植面积超过 10 万亩的品种有台农一号、桂热芒 82 号（桂七）、金煌和桂热芒 10 号。

上市时间：6 月下旬至 8 月底。

分布概况：百色市的芒果种植面积达 133 万亩，产量超过 90 万 t，占广西总量的 85% 左右。而这其中的 80% 集中在右江河谷，右江河谷已成为全国连片规模最大的芒果产区。

区域品牌："百色芒果"，中国国家地理标志产品；2019 年入选中国农业品牌目录；2020 年入选中欧地理标志首批保护清单。品牌价值评估达 173 亿元。

2.3.2.4 云南西南—云南南—云南中元江流域芒果优势产业带

种植面积约 137 万亩。

品种结构：云南芒果品种多样，主要品种有三年芒、象牙芒、缅甸 3 号、凯特、圣心芒、马切芒、台农一号、金煌、贵妃等。

上市时间：5—11 月。

分布概况：云南省是全国鲜食芒果供应期最长的省份，现形成了怒江流域、红河流域和金沙江流域早、中、晚熟 3 个特色芒果优势产业带。云南省种植芒果的县市有 71 个，面积最大的县市是华坪 42 万亩，其次是红河县 11 万亩，元阳、河口、弥勒、新平、隆阳、永德等地也超过 10 万亩。

区域品牌："华坪芒果"，中国国家地理标志产品，获得了国家级、省级特色农产品优势区、全国名优果品区域公用品牌、云南省十大名果等称号。

2.3.2.5 金沙江干热河谷流域晚熟芒果优势产业带

种植面积超过 100 万亩。

品种结构：以晚熟芒果为主，早中晚熟搭配。其中，凯特、吉禄、热品 10 号等攀枝花晚熟芒果占全年产量的 80% 左右。

上市时间：6—12 月，集中上市时间是 8—11 月（图 2-2）。

分布概况：攀枝花芒果种植面积 103 万亩，产量 54.5 万 t。在攀枝花种芒果有一个得天独厚的优势，就是芒果在这里比在海南、广西等其他产区晚熟。规模化芒果种植基地集中在仁和、米易、盐边 3 个农业县（区）。

区域品牌："攀枝花芒果"，中国农产品地理标志，2020 年入选中欧地理标志第二批保护名单。

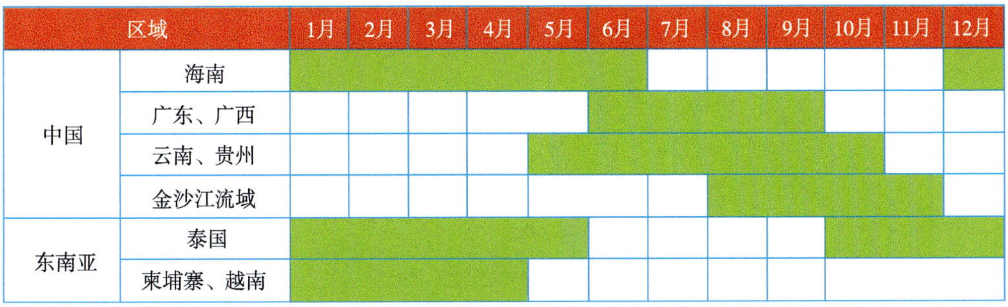

区域		1月	2月	3月	4月	5月	6月	7月	8月	9月	10月	11月	12月
中国	海南	●	●	●	●	●	●						●
	广东、广西					●	●	●					
	云南、贵州				●	●	●	●	●				
	金沙江流域							●	●	●	●	●	
东南亚	泰国		●	●	●	●	●						
	柬埔寨、越南			●	●	●							

图 2–2　主要芒果产区芒果上市时间

近年来，我国芒果产区呈现的特点是"早果越来越早，晚果越来越晚"，导致销售季节错开的两个产区，重叠度越来越高。一定市场范围内，芒果市场竞争压力大，抵御风险的能力较弱，芒果的价格也会受到影响。

芒果种植管理技术

芒果具有丰富的营养价值和药用价值,深受大众喜爱,市场前景广阔。芒果产业也是热带亚热带地区的特色产业,优质的管理是实现芒果高产、优质、高效生产的关键。

3.1 芒果种苗及其繁殖技术

芒果的繁殖方法可分为有性繁殖和无性繁殖两种。有性繁殖即用种子播种繁殖,所繁殖的种苗一般习称"实生苗"。无性繁殖包括嫁接、空中压条及扦插等,一般以嫁接法较为常用。

3.1.1 有性繁殖

3.1.1.1 苗圃的选择

应选择地势较为平坦、向阳、避风的平地或缓坡地。土壤以土层深厚且排水良好富含有机质的砂质壤土为宜。芒果苗圃在整畦时宜注意排水,因为在育苗期的幼苗怕浸水,连续多日浸水易导致幼苗发育不良或死亡。整畦时应视土壤肥力状况,酌情施腐熟的有机质肥料及过磷酸钙做基肥。

3.1.1.2 催芽、播种

种子的选择以种核饱满为佳,播种时可选用自鲜食后至 20 d 左右的种子,稍加洗净残肉阴干即可。在播种前须先行剥壳处理,因芒果种子具有坚硬的外种皮(外壳)而影响发芽,先除去外种皮及覆土 2 cm 可有效提高发芽率(图 3–1)。

种子在室外风吹日晒 7 d 左右即丧失发芽能力,若稍加洗净阴干放置于半湿润的沙土中储藏发芽，则可保存约 30 d。播种时，种仁宜直立，采取三角形定植法，种脐朝下，不可平放播种，否则会引起萌芽不整齐。种子种下后约 7 d 可发芽。播种行株距以 10 cm×15 cm 为宜，覆土约 2 cm，播种后每隔 1~2 d 浇水一次，保持苗床湿润。种子发芽后至抽第二次梢时，可施追肥一次或以 2% 尿素水溶液浇灌一次，种子苗 6~9 个月后即可做为标准的嫁接砧木。

 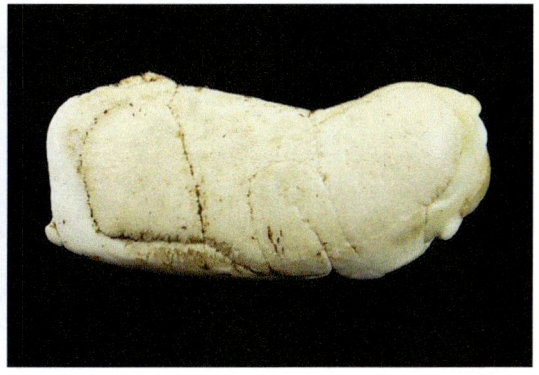

图 3–1　芒果果核与胚

3.1.2　无性繁殖

芒果嫁接可供大量繁殖优良种苗，以及用于已定植果园更新品种。嫁接属无性繁殖，不易变异，能保持母本的优良性状，在育种上只要育成一个新的优良品种，即可用嫁接法大量繁殖，如金煌、贵妃等现有品种。嫁接的方法有多种，一般采用切接法较为简易，而对于如台农一号等嫁接不易成活的品种，亦采用靠接法。

3.1.2.1　嫁接时间

在不同的嫁接时期中，以 3 月的接活率最高，9 月次之，6 月及 12 月最差。3 月正是春天来临，温度回升气候温和，也是芒果萌芽生长的开始，植株在嫁接后愈合力强，因此嫁接成活率高，大部分苗圃及果农都选在此时嫁接。6 月高温多湿，降雨频繁，病虫害多，嫁接时湿气重，套上塑胶袋后，砧木与接穗接口易发霉，而导致嫁接愈合组织无法产生或产生后无法愈合，因此雨季并不适合嫁接，如嫁接后未遇雨季，则接活率和 3 月相似。9 月为秋高气爽的季节，南方地区为雨季末期，此时为良好的嫁接时期，芒果植株的秋梢也正在此时抽出，但其接活率仍稍逊于 3 月。12 月为冬季的开始，平均气温将降至 18℃ 以下，大部分植株生长迟缓，不宜嫁接。综上所述，理想的嫁接时期约在 2 月下旬起至 5 月雨季来临前，而以 3 月为最佳嫁接时期，错过此时就须等到雨季结束，到 9—10 月再嫁接，而在第一波寒流来临前应停止作业。

3.1.2.2 砧木选择

优良的砧木应具备的特点：抗病虫害能力强；种子来源经济方便，繁殖生长快速；嫁接亲和力强；能适应当地的气候环境；嫁接后果实品质不会劣变及产量不会降低等。

嫁接砧木以 1~2 年生茎粗 1 cm 的苗木为最佳，因其发育快速，嫁接成活率高。如果在 7 月播种，生长良好的幼苗到翌年 3 月时生育期约 250 d，即可用作砧木。

3.1.2.3 接穗的剪取

一般接穗应选择生长旺盛，无病虫害的健壮植株，接穗以当年生成熟或半成熟直立枝条较佳，倒垂或横枝均不适宜。接穗的成熟度须和砧木配合，如以 1~2 年生砧木供嫁接时，接穗应选较嫩或半成熟的枝条。如砧木为多年生时，接穗应剪自成熟的枝条，不宜剪自植株尚在幼年期的枝条，也不宜剪自植株正值开花、结果或采收后 30 d 内的枝条，因为植株如果正值生殖生长或营养不足，剪取的接穗不易接活。

采穗的时间以上午为最佳，接穗采后，如台农一号须在 1 h 内嫁接，若接穗有乳胶流出时，嫁接成活率高，反之则低。采穗后应立即剪去叶片，稍留 0.5 cm 的叶柄，避免伤及潜芽，并应立即嫁接。如需运输应以塑胶袋包扎，储藏时可放于冰箱中低温储藏，储藏时间约 5 d，放置愈久，接活率愈低。

嫁接成活率低的品种可采用催芽或剪叶处理来提高接活率。在嫁接前 7 d，将拟供做接穗用枝条的顶端剪断，枝条有顶芽优势，约 7 d 后顶端的潜芽会渐渐突起，此时剪下供做接穗将会提高接活率。剪叶处理为在嫁接前 7 d，将提供接穗枝条的叶片剪除，仅留约 1 cm 的叶柄，以此处理后也能提高接活率，但若枝条喷波尔多液后剪下用作接穗，嫁接则不易成活。

3.1.2.4 嫁接方法

嫁接前应准备：接穗、砧木、枝剪、锯子、切接刀、透明塑胶袋、旧报纸、绳子、标签及笔（记录嫁接日期及品种）等。

在苗圃的嫁接，可将砧木离地面 30~40 cm 处剪下，以切接刀将嫁接口削平后，沿木质部及韧皮部的中央切下约 3 cm，接穗剪成每段 5~6 cm，以顶端留有两芽为准，后将接穗较平的一边，直削约 3 cm，深及木质部即可，对面的另一边斜削约 0.8 cm 后将接穗插入砧木中，注意一边须靠紧形成层（图 3-2），再以塑胶绳束紧，套上透明塑胶袋，外面以旧报纸包扎即告完成。嫁接后在温度高时约 21 d 可成活，温度低时约 28 d 可拆开纸袋，检视是否成活，如果没有接活，接穗会萎缩干枯或发霉，当接活时，接穗仍保持原来的颜色，且自潜芽处稍为隆起，而后长出新芽、新梢。如拆封时接穗尚未长出新梢，应将塑胶袋内的水分拭去重新套上，再包扎报纸，于 5~7 d 后再重新检查一次。当接穗长出新梢后，可取下或剪开塑胶袋，外面包扎旧报纸后，任其生长。

成年植株的嫁接和苗木嫁接的方法及原理相同，唯一不同的是接穗的选择应以较成熟的枝条为佳，且苗木以一株接一穗，而成年植株则视枝干大小不同可接 1~3 穗。成年植株的嫁接一般以更新优良品种为目的，因此原有的品种如果未经嫁接时，可锯至离地面 60 cm 左右再行嫁接。如已嫁接，再次更新品种时，则应锯至原来砧木的地方再行嫁接，若嫁接失败，可在砧木处培养新梢，在不同方向预留约 3 支，待枝梢成熟后接在新梢上，虽日后的发育比接在大砧木上稍慢，但较易接活。

图 3-2　芒果嫁接技术

3.1.3　培育砧木苗

3.1.3.1　苗圃地的选择与整地

选择靠近水源、避风而冷空气不易沉积、土层深厚、有机质含量丰富、排水良好的壤土或沙壤土开辟苗圃，周围 1 km 内无芒果园和芒果老树，以减少病虫害，有利于培育无病虫害的苗木。

苗圃地要三犁三耙，保持土壤疏松，开好排灌沟，地下水位高或地势低者应起高畦，畦长 10 m，宽 0.8~1 m，高 0.2 m，畦间距离 0.4~0.5 m。苗圃必须施足基肥后浅翻，使土肥混合均匀，以保证苗木生长所需的营养。苗圃地可每公顷施腐熟畜粪或堆肥 4 500~6 000 kg。

3.1.3.2　种子的采集和处理

用作砧木种子的果实要选自生长健壮的母树，果实充分成熟且饱满新鲜。在生产上可采用罐头厂当天加工后留下的种核，也可购买大量饱满果实堆烂后取出种子。一般本地土芒出芽率高、出芽整齐及苗木健壮，最适用于砧木苗的培育。种子取出后不宜久存，洗去果肉后晾干，在 5 d 内播种，超过 7 d 后发芽率严重下降。长途运输种子时可用湿椰糠、炭粉或湿河沙等储藏。作砧木用的种子应选用与当地推广品种亲和力强的品种，一般大叶品种的种子（含优良栽培种）都不宜选作砧木用。

3.1.3.3 剥壳催芽

芒果种子外木质硬壳妨碍种子发芽，直接用种核播种发芽率低，种苗弯曲、畸形比例大，应采取剥壳催芽的方法以提高出圃率，培育健壮砧木苗。实践证明，剥壳催芽发芽率达90%以上，未剥壳的种子发芽率仅为40%~60%或更低。

3.1.3.4 分床移植

苗床可采用双行式或四行式，双行式行距20~30 cm，株距20 cm，畦面宽50 cm，四行式行距20~30 cm，株距20 cm，畦面宽1 m，移苗时用竹签小心将幼苗连种仁全根挖出，移于苗床，行距22~25 cm，株距15~20 cm，在移苗过程中注意不要损伤幼苗根系。主根过长可适当剪短，但根长不应小于10 cm。如用营养袋育苗，则把苗移入盛营养土的袋中［1/3细沙、1/3泥土、1/3腐熟杀菌后的有机肥混合后装袋，袋大小以（22~24）cm×30 cm为宜］，将幼苗按每袋1株苗移入袋内，酌情适量盖土后淋足水即可。

3.1.3.5 砧木苗管理

淋水施肥：移植后及时淋水，每天淋水1次，保持苗床湿润。为保证苗木生长所需养分，小苗抽新梢时开始追肥，可施（1:5）~（1:4）的粪水或0.5%~1%的硫酸铵或尿素水溶液，以后每抽1~2次梢追肥一次。

遮阴：移苗时宜选阴天，生长未超过3个月的幼苗组织幼嫩，应在苗圃上方以50%~60%的黑色遮光网搭荫棚，以保护苗木。

防治病害：苗期主要的病害有炭疽病、叶斑病等，可用多菌灵和波尔多液防治。

3.1.4 培育嫁接苗

3.1.4.1 嫁接时期

一般以每年3—5月及9—10月为最适宜嫁接期，此时气温稳定，湿度不大，苗木处于生长旺季前、后期，成活率高。

3.1.4.2 接穗的选择与储存

接穗须采自接穗圃或长势健壮、无病虫、品种纯正的母树，选健壮、充实、叶芽多且芽眼饱满的1~2年生枝作接穗。采下的接穗剪去叶片，注意不能伤芽并及时嫁接。如需储存运输3 d以上，可用湿椰糠保存，或直接放入聚乙烯薄膜袋中，然后保湿装箱。

3.1.4.3 嫁接方法

（1）切接法

其优点是能利用较幼嫩的接穗；不受物候期和剥皮难易的影响，只要温度条件允许，任何时候都可以嫁接，且成活后抽芽与成苗快。具体操作如下。

①削砧木：在离地 25 cm 左右处将砧木剪断，剪口应向平直一侧面稍为倾斜，然后在斜切面下部切一长约 2 cm 的切口，深度以削去少许木质部为宜，然后将带少许木质部的皮层切去 2/3。②削接穗：用与砧木粗度相近的枝条（宜小不宜大），选 1~2 个芽作一段接穗切下，将其一侧从上到下滑削一切口，长与砧木切口相同，深至削去少许木质部为宜，在切口背面末端削成 45° 斜口。③安放接穗及绑扎：将接穗下端插入砧木切位，使接穗与砧木切口的皮层相吻合，然后用超薄薄膜将吻合部分先绑紧，再将接穗密包扎好。用超薄薄膜无须解绑，芽可穿透薄膜而出。操作时要求刀利、动作快、保持形成层完整、干净、对接准确、缚扎紧密。

（2）芽接法

其优点是操作简单，易学，利用接穗较经济，接口愈合快，补接方便，但砧木和接穗必须易剥皮才能芽接。具体操作如下。

①开芽接位：在砧木离地 20 cm 高处选表皮光滑、无叶节的枝段用刀尖刻一宽约 1 cm、长约 2.5 cm 的长方形口子，长边与树干方向平行，深达形成层，开位后自上而下将皮层剥开并切除大部分，留下端小部分以承接接芽。②削芽片：宜选用芽眼饱满的腋芽或密节芽，在芽眼上方或下方削取 1 cm 以上的位置削取长 3~4 cm、宽 1.2~1.5 cm 的芽片，按砧木接口宽度修平芽片两边，小心将皮层和木质部分离，注意不要损伤形成层，最后把芽片削成比接口稍小的长方形。③安放芽片及绑扎：将剥下的芽片顺向、端正平放在芽接位的中部，然后用弹性好的塑料薄膜带用力均匀绑扎紧至完全密封为止（图3-3）。

图3-3 芒果苗嫁接

3.1.4.4 嫁接苗管理

（1）解绑及补接

用枝接法进行嫁接的，芽抽出后让其穿破薄膜，待芽长出1次梢时再解绑。解绑时用刀片在接口背面一侧将塑料带割断，不要靠近接口，不要割破皮层。用芽接法进行嫁接的待嫁接3周后，用嫁接刀在接口处切开绑带，5 d后检查芽片是否成活，若芽片保持绿色，则可在接口上方1 cm处剪去砧顶，若芽片变为褐色，表明芽片已坏死，要在接口下部补接。

（2）除砧木芽

嫁接后，从砧木基部或剪砧后剪口处易长出萌芽，要及时多次去除，以免和接芽争夺养分，影响接芽生长。

（3）肥水管理及病虫害防治

嫁接苗剪砧后要及时淋水以促进接芽抽穗，每次抽梢时，施化肥1次或淋水肥1次，做到勤施薄施。初抽生的接芽组织幼嫩，易受病虫害侵扰，因此每次抽梢时应喷药加以防治。

3.1.4.5 苗木出圃

（1）苗木出圃标准

苗木出圃前应对苗木品种、数量、质量进行核实标记，制订出苗出圃计划。在生产上要求达到品种纯正，砧穗接口愈合良好，根系发达，须根多，苗高约80 cm，至少有3蓬老熟的枝梢，无病虫害（图3-4）。

图3-4 准备出圃的芒果嫁接苗

（2）起苗

芒果起苗分带土起苗和不带土起苗。带土起苗应在晴天进行，可利用起苗器等工具，保留直径15~18 cm的土团，起苗后剪去1/3叶片，然后用稻草包扎。不带土起苗要在起苗前1 d灌透水，否则起苗时极易损伤根系，也增加挖苗的困难。

3.1.5 芒果园种植选址与要求

3.1.5.1 园地选择

芒果园选择应以排水良好、土层深厚（2 m以上）、土壤松软肥沃、有机质含量丰富的沙壤土为佳，土壤的透气条件好，不积水，另外地下水位要保证在1 m以下（图3-5）。

图3-5 芒果标准化种植果园

3.1.5.2 果园开垦、规划及种植准备

开垦是果园规划中一项基础性工作，包括砍伐、清地、翻耕、平整土地、开壕沟或梯口、定标挖种植穴或开种植沟等。

在种植前1~2个月挖好种植穴，并进行穴土回沟，先把表土置于最下层，接着放杂草、枝叶和0.3 kg石灰混合，再回一层表土，放塘泥、土杂肥30 kg和0.2 kg的磷肥，再回盖一层心土填充至地面20 cm时，放入经混合堆沤后的禽畜粪10~15 kg，钙镁磷肥1~1.5 kg，复合肥0.2 kg，再加回土拌匀填至高出地面15 cm，然后盖土10~15 cm做成上口径约60 cm的土盘，这样待20 d左右土盘充分下沉后应高于地面，以防种植穴下沉，埋没苗木根颈，且易积水，不利于生长。上述工作完成后，便可以开始种植。

3.1.5.3 种植规格

（1）定植密度和时间

芒果最适宜的定植季节一般为3—5月，其次是9—11月。栽植株行距可根据品种特性、园地土壤肥力和管理水平及机械化程度而定，一般株距3.5 m，行距4.5 m，亩植33~42株；台农一号、爱文、凯特树势中等，树体较矮，树冠较小，可适当密植。

（2）定植方法

在定植前要进行苗木处理，首先要消毒，外地调入苗木栽前可用3~5 Bé的石硫合剂药液喷布苗木进行消毒，同时要进行分级和修剪。植穴施基肥15 d后才能定植。定植时，先将营养土层挖一"V"字形穴，深30~40 cm（按苗木根长短而定），以芽接苗根颈高出地面5 cm为准，将苗放下。若是裸根苗，须将根系分层自然伸展，分层盖回表土，轻轻压实，再回覆营养土，用脚压实；若用营养袋苗，放下植穴后先将袋撕破或取出，后直接回覆营养土，轻轻压实，上面再盖上一层10~15 cm心土，形成高出地面10~15 cm的窝形土堆。定植后灌透水，水渗下后立即培土以防水分蒸发和苗木动摇，然后在窝形土堆上用枯枝树叶加以覆盖，以利保水。

3.2 芒果栽培管理技术

3.2.1 树型管理

3.2.1.1 整形修剪

整形修剪指通过整形和修剪使芒果具有良好的树体结构，造就一个充分利用空间和阳光，调节生长与发育节奏便于管理的树型。

整形修剪应按不同品种的特性、种植密度、水肥管理等因素来考虑确定树形、修剪方法和强度；矮化树形，一般将树体控制在2.5~3 m，树体结构要层次分明，充分利用空间和阳光，又具有通透性；修剪整形必须与水肥管理密切配合才能达到预期效果。

3.2.1.2 整形技术

芒果树的树体结构：芒果树的地上部分包括主干和树冠两部分，树冠由中心干、主枝、侧枝和枝组构成，其中中心干、主枝和侧枝构成树冠的骨架，统称骨干枝。

主干：从地面根茎处至第一分枝处的茎干称为主干。一般品种定干高度为50~80 cm，分枝极性强的品种如紫花芒、粤西1号等品种，定干高度可低于50 cm。

中心干：在主干以上的骨干枝称为中心干，是结果枝组的着生处之一，其留取的条数一般也与品种有关。分枝极性强的品种一般留取 1 条，弱性品种留 2 条。

主枝：着生在中心干上的骨干枝称为主枝，是果树树冠的主要骨架，侧枝的着生处，树形大，主枝多；树形小，主枝少。芒果主枝每株树 5~7 枝，同层不多于 3 枝。

侧枝：着生在主枝上的枝条为侧枝，它是叶片着生和开花结果的主要部分，也是果树树冠的骨架之一和各类枝组的着生处，整形时要尽量多留。芒果侧枝第一结果年多属结果母枝，收果后萌发的新枝成为第二年结果母枝，全树一般保留 30~40 条。

3.2.1.3 修剪方法

短截：又叫短剪，即剪去枝梢的一部分。它可增加分枝，促进枝梢生长和更新复壮，改变不同枝梢间顶端的部位，从而改变顶端优势的部位，调节主枝的平衡。

疏剪：又叫疏删，即将枝梢从基部疏除。它的作用为减少分枝，利于树冠内的通风透光，促进花芽分化与结果。

摘心：摘心指新梢长到一定程度时将其顶端最嫩部分用手摘去或用剪子剪掉，促使芽充实和形成花芽，提早结果，并提高坐果率。在芒果幼树整形时常用摘心来促进分枝。

除萌：抹除嫩芽称为除萌或抹芽。

环剥：是将枝干的韧皮部剥去一环。在芒果生产上对营养生长过旺的树可采用环剥来达到促进花芽分化、利于花芽形成的目的。环剥的时间和宽度要适当，一般环剥宽度为被剥枝直径的 1/5~1/3，否则达不到预期目的。

另外，弯枝、扭枝、刻伤、断根、去叶等也是修剪措施，需要时可适当采用。

3.2.1.4 培养丰产树形

芒果不同的品种（或类型）有不同的树形，目前商业化栽植的芒果树主要有疏散分层形、自然圆头形、自然扇形三类，整形时要根据不同的树形加以区别对待。

1）结果树的修剪

（1）采果后修剪

在果实采收后的 8—9 月进行，这一次修剪为全年修剪最大的一次，目的是调整树冠永久性骨干枝的数量和着生角度，使其分布均匀。方法以短截结果母枝为主，并适当剪除过密枝、过多主枝，回缩冠间和冠内的交叉枝、重叠枝，剪去下垂枝和病虫枝。

（2）生长期修剪

生长期的修剪包括秋梢修剪、春夏梢修剪等。主要采取抹芽、疏梢和短截的方法。秋梢的修剪：经采果后修剪抽出的秋梢，根据空间位置保留 1~3 条，其余的抹去，留下的枝梢长至 18~20 cm 时短截，促发第二次抽秋梢，末次秋梢留取长 18~20 cm，中等粗壮的枝条作为结果母枝，其余的抹去。春夏梢修剪：在开花不足的小年，春夏梢大量抽生旺长，新梢对养分抢夺剧烈，导致花序抽生纤弱，坐果率不高，因此，对开花量不足 50% 的树视情况疏枝抹芽，以保证足够的营养促使花序正常抽生。此次修剪在春梢抽出后，即第二次生理落果后（果有鸡蛋大时）进行，目的为了将未结果的春梢培养为来年的结果枝，使大量的养分能完全供给果实的发育。

（3）培养结果母枝

芒果通常是顶花芽，末级梢只要适时老熟均可成为结果母枝开花结果。在我国大部分芒果产区，丰产树很少抽生春梢和夏梢，因此一般都以果实采收后抽出的秋梢或早冬梢作为结果母枝。促发适时停止生长的秋梢或早冬梢是培养优良结果母枝的关键（图3-6）。

图 3-6　丰产树形培养

2）老树更新

树冠更新宜在春季萌芽前进行。伤口用1%硫酸铜等药剂消毒，并用接蜡等保护剂涂封，以防病害，减少蒸发，促进愈合和剪口梢生长。主枝更新时，要用石灰乳喷射树冠骨干枝，以免发生日烧病。更新枝抽生时，宜注意防风和整形修剪，对树干下部抽生萌蘖的，要及时去除。

3.2.2　土肥管理

3.2.2.1　树盘管理

芒果定植后的幼树期要注意树盘的管理，做到保湿防鼠。一般用稻草等覆盖树盘，保持一定的湿度。

3.2.2.2 果园生草与覆盖

新植或植树后 1~2 年的芒果园，可在果树行间播种豆科、禾本科草种进行树盘覆盖，如柱花草、假花生、毛蔓豆、无刺含羞草、热带苜蓿等。覆盖作物长成后，花期或始花期进行割锄，然后对树盘实行清耕覆盖，能起到铲除杂草、调节地温、保墒、改良土壤结构，促进树盘根系微生物活动的作用，覆盖厚度以 15~25 cm 为宜。对雨水分布不均、冬春干旱明显的雷州半岛及海南西南部地区的果园，具有明显的保水降温效果。此外，果园间种花生、黄豆，采用秸秆覆盖，经济效益和生态效益俱佳。

3.2.2.3 扩穴改土

除雨天、高温干旱天气和盛花期等时期外，全年大多数时间都可进行扩穴改土，但以秋冬为宜。

扩穴改土的方法：改土坑的形状和位置依定植穴的形状而定。如开壕沟定植的则每年轮流在定植壕沟的一侧开壕沟改土；如定植穴是圆的，则每年轮换在株间或行间开两个 1/4 周弧形改土；如定植穴是正方形或长方形，则每年轮流在株间或行间各挖一个长方形坑，进行"井"字形扩穴，扩穴深度大约为 40 cm，宽度至少 60 cm，坑的长度要随树龄而逐渐增加。

3.2.3 营养与施肥管理

3.2.3.1 芒果的营养特性

1）芒果养分需要量

一般亩产 1 061 kg 鲜果，树体需要从土壤中吸收氮 104 kg、五氧化二磷 27.5 kg、氧化钾 119 kg、钙 88 kg、镁 47 kg、锰 871 g、硼 174 g、锌 375 g、铜 435 g、铁 976 g，主要养分吸收比例为氮：五氧化二磷：氧化钾：钙：镁 =1：0.26：1.14：0.85：0.45。在我国，每收获 18 668 kg/hm^2 的紫花芒，果实养分吸收量为氮 22.4 kg/hm^2，五氧化二磷 9.0 kg/hm^2，氧化钾 44.7 kg/hm^2，钙 3.2 kg/hm^2，镁 3.0 kg/hm^2，硫 2.3 kg/hm^2，养分吸收比例为氮：五氧化二磷：氧化钾：钙：镁：硫 =1：0.40：2.00：0.14：0.13：0.10。而每收获 1 000 kg 鸡蛋芒，鲜果所带走的养分（平均值）分别为氮 693 g，磷 231 g，钾 1 575 g，钙 225 g，镁 212 g，氮：磷：钾：钙：镁 =1：0.33：2.27：0.32：0.31。

2）不同生长期矿质养分的变化规律

一般随着芒果秋梢的生长和老熟，叶片中的氮、磷、钾、钙、镁、锌和硼含量逐渐升高，至秋梢成熟时达到最高值。当芒果进入开花期，叶片养分向花穗转移，开花期 1 个多月叶片养分大幅度下降，以氮、磷、硼、锌、锰和铁等养分消耗较多。至幼果期叶片养分下降趋缓，到第二次生理落果期（果实玻璃珠大小）时，果实氮、磷和硼含量明显增加，而钾、钙、镁浓度增加相对较小。至果实快速膨大期时，氮、磷和硼浓度降低，而钾、钙和镁则明显增高，达到最高值。果实成熟后，矿质养分含量均显著下降。

3）芒果树的营养诊断、施肥反应及主要矿质营养缺乏和过剩的症状及其矫正

（1）氮（N）

氮是控制植物生长的主要养分元素，它存在于植物的多种化合物中，在植物生长最活跃的部分所需浓度最高。幼嫩叶片、花、根端都利用了氮。在果园中，最可能缺乏的就是氮。缺氮时叶片呈黄色，症状首先出现于老叶，尔后波及顶部叶片，随着时间的推移，黄叶的叶尖和叶缘出现坏死的斑点，落叶并伴有严重的生理落果，枝梢细弱，花芽小，果实小而色浓。长期缺氮素会降低植物的抗逆性。我国沙质土果园一般年施氮量以 400 g/ 株为宜，黏质土果园以 600 g/ 株为宜。

（2）磷（P）

磷是始花、坐果、根系发育的重要养分，磷在土壤中移动缓慢，在下雨前施用效果较好。芒果缺磷症状首先在老叶上出现，叶脉间有坏死的褐色斑点，最后布满全叶，缺磷叶片早期脱落。严重缺磷时，树体生长迟缓，分枝少，叶小，花芽分化不良，果实成熟晚，产量下降。在我国，结果树年施磷量一般以 150 g 五氧化二磷 / 株为宜，磷过量时会影响树体对氮、钾、铁、锌、铜的吸收，并易造成冬梢萌发及夏梢大量生长，对坐果产生不良影响。

（3）钾（K）

芒果对钾的需求量很大，钾对于秋梢的生长、果实的发育和果实质量有很重要的作用。缺钾植株的叶片小而薄，先是在老叶上出现不规则的黄色小斑点，尔后枯斑沿着叶缘在叶脉间扩展，缺钾初期坏死仅限于叶缘，严重时扩展至全叶，灼枯叶片可在树上存留数月，着色不良。结果树钾肥年施用量，砂质土芒果园氧化钾以 500~600 g/ 株为宜，黏质土果园以 600~700 g/ 株为宜。施钾量过高易引起树体营养不平衡，使叶片钙和镁含量降低，增产效果反而降低。

（4）钙（Ca）

钙的作用是调节芒果生长环境，提高芒果抗病力，减少芒果生理障碍。缺钙时芒果叶片呈黄绿色，且顶部叶片先黄化，严重时，老叶除近叶尖和叶基的小面积外，沿叶缘全部都带褐色的伤状，叶片卷曲，边缘皱缩，易破裂，顶芽干枯，花朵萎缩。根系受害突出，新根短粗，弯曲，尖端不久褐变枯死。开花及幼果期缺钙，就会授粉不良，结实少，落果严重。果实成熟期缺钙，外观表现为果实尖端软化，褐变，尤以果顶腹部果肉发病严重，即所谓软鼻病。在酸性土壤上施石灰，硝酸钙或氯化钙可以作为缺钙症的应对措施，此外，果实采收后浸钙也有一定效果。结果树钙肥年施用量：一般黏质土石灰用量以 0.95~1.26 kg 生石灰 / 株，沙质土以 0.62~0.87 kg 生石灰 / 株为宜。

（5）镁（Mg）

镁是植物叶绿素进行光合作用的重要元素，对芒果秋梢生长、叶片光合作用以及坐果有重要影响。缺镁时植株生长受到抑制，叶片变短略变宽，成熟叶片的叶尖及叶缘坏死，主脉呈暗绿色，在暗绿区与坏死区之间的叶肉淡黄色，叶脉仍显绿，失绿呈鱼骨状。叶片提早脱落，果实成熟推迟。缺镁症从老叶开始，逐渐向幼嫩部分扩展。缺镁时，应避免过多施用钾肥，在生长旺盛季节可用 0.5% 的硫酸镁进行叶面喷施。芒果结果树镁肥年施用量一般为硫酸镁 1 kg/株左右。

（6）硫（S）

硫是叶绿素形成的必要元素，也是组成蛋白质不可缺少的元素。缺硫症状出现较晚而缓慢，症状为叶片一老熟就沿叶缘出现坏死，在 15~20 d 内整片叶变为褐灰色，叶片变得质脆，易早期脱落。一般典型缺硫症状较少见。结果树硫肥年施用量为 40~80 g 硫/株。

（7）微量元素

主要包括硼、锌、铁、铜、钼、锰等，为芒果生长发育所必需。我国芒果产区养分水平较低的微量元素主要有硼和锌。芒果缺硼幼梢节间变短，顶芽易枯死，叶片厚而质脆，不能坐果；缺锌则出现小叶病，簇生小叶发生在顶端幼嫩部分，枝条下部叶片叶脉间失绿，叶早落，果实小。花期喷施 0.3% 的硫酸锌溶液和坐果后喷施 0.5% 的硼砂溶液，每隔 20 d 一次，连续两次，可明显缓解缺锌和缺硼症状。

3.2.3.2 芒果的施肥量、施肥时期和方法

1）幼龄树施肥

对于 1~3 龄幼树，为了促使幼龄芒果树快速生长，尽快形成发达根群，肥料的施用应与枝梢生长物候相结合。芒果种植后，沙质土以一梢二肥、黏质土以一梢一肥较合理。肥料应以速效氮和磷肥为主，并注意培肥地力。为使芒果形成强大的根群，每年可施用 3 次有机肥、2 次石灰。有机肥分别在春梢、夏梢和秋梢萌动前施用，石灰分别在春梢和秋梢萌动前施用。具体施肥量如表 3-1 所示。肥料施用方法：化学肥料可兑水淋施，亦可开半环沟结合灌水施用，有机肥和石灰可开环沟，或结合扩穴施用。

表 3-1 芒果幼龄树施肥用量

单位：kg/（株·年）

树龄（年）	养分用量				肥料用量					
	N	P_2O_5	K_2O	Mg	尿素	过磷酸钙	氯化钾	硫酸镁	有机肥	石灰
1	0.075	0.075	0.06	0.01	0.16	0.56	0.1	0.1	60	1
2~3	0.15~0.2	0.2	0.2	0.02	0.33~0.43	1.5	0.33	0.2	60	1

2）结果树施肥

芒果嫁接苗定植后第四年开始进入结果期，此时期施肥应考虑到开花结果、果实发育的不同阶段而进行营养补充。根据芒果树养分周年变化规律，按结果树的物候可分为4个施肥时期。

（1）果后肥

在芒果收获后第1次修剪前进行，果实和枝梢修剪带走了大量养分，树体营养量消耗较大，此时宜施果后肥进行补充，以供植株迅速恢复生势。果后肥要及早施用，一般在修剪前5~10 d时施用。施肥以有机肥和氮、磷为主，另外结合清园配施石灰。

（2）攻秋梢肥

在我国南方气候条件下，一般在8—9月开始抽秋梢，秋梢生长期生物量占全生育期生物量的30%~50%，此时植株积累大量光合产物和矿质养分，为花芽分化做物质上的准备。抽秋梢时应追一次肥，施肥时以氮、钾肥主，配施有机肥。此外，若修剪迟，秋梢抽生较缓，则可结合叶面喷施尿素（1%~2%）、硝酸钙（0.5%~1.0%）、磷酸二氢钾（0.2%~0.5%）等以加速秋梢生长。

（3）催花肥

催花肥的施肥时期在开花前半个月左右，海南、云南大部分热带地区一般为1月，其余地区为2月或3月。此期间是芒果抽穗开花季节，对氮、磷需求较多，在抽穗前期使用速效氮、磷肥，配合硼（0.2%硼砂）、锌（0.2%硫酸锌）等微量元素进行叶面喷施，以提供抽穗期开花的营养需要，提高花质，促进开花结果。

（4）壮果肥

施肥时期在果实玻璃珠大时（果实膨大前），此时由于幼果进入迅速生长期，大量幼果需从叶片争夺营养，因此需增施一次肥料用于补充营养，促进果实膨大，减少落果。施肥以磷、钾肥为主，配施氮肥。此外，还可于此时每隔15 d喷赤霉素（2.5 mg/L）、细胞分裂素（5 mg/L）等物质组成的保果剂，共喷2次。

3.2.4 芒果园清园及其管理

3.2.4.1 清园目的

一是修剪枝条，培养良好的适合自己果园的树冠、树形；二是适当矮化树体，便于日后的管理；三是清理芒果园的病菌残体，降低果园病菌基数；四是促梢整齐，培养健壮的枝梢，为下半年反季节生产培养良好的结果母枝（图3–7）。

3.2.4.2 采果后清园主要建议

1）养树

（1）存在问题

土壤酸化与板结：芒果目前主要种植在红黄沙壤上，多年来为了追求产量，重施化肥，轻有机质的补充，土壤中酸性物质不断积累，加重与加速土壤酸性，土壤贫瘠，土壤有害金属活化，土壤有害微生物特别是寄生真菌增加，土传病害严重，根系生长受到抑制，导致芒果树体长势衰弱与营养失调，严重的甚至导致植株死亡。

土壤营养失衡：土壤长期被淋溶与侵蚀，有机质大量流失，土壤中微量元素如钙、镁、锌、硼、钼不足或缺乏，钾和硅元素缺失，磷被固定，各类元素之间比例失调，

图3-7 芒果园清园

有益元素缺失，并导致土壤元素之间出现拮抗作用，有肥而无用。

激素盲目施用：目前部分芒果产区在控梢、促花、果实膨大过程中使用了大量激素，如多效唑、乙烯利等，导致树体透支，营养补充不均衡，各类生理症状表现突出，如缺素症、树体流胶、长势和萌梢能力差、枝梢质量差、开花结果能力弱、果实品质差等，各种问题层出不穷。

（2）建议措施

改好土、促好根、保好叶、促好花，保好果，这是增产增收的前提。

改好土：土是根本，是一切作物正常生长的基础，只有土壤疏松，通风透气，才利于根系生长与肥水吸收，土壤过酸或过碱都会抑制根系生长，引起根系发育不良，根系老化，新根群少，严重的造成死根，肥水吸收不畅，植株长势弱。

促好根：树体的营养主要靠根系通过土壤吸收，叶面肥只是起到补充作用。若根系生长不好，肥水难以被吸收、运输、转化以供树体生长所需，并且营养失衡。尽管树体能萌梢，但会表现出萌梢困难，不整齐，萌梢瘦弱，同时枝梢短簇，容易出现相应的缺素症状，对培养健壮的枝梢与结果母枝造成很大的影响，进而影响到花果的质量。同时树体由于营养失衡，导致树体流胶现象加重，树体弱化，更容易引起树体感染细菌性角斑病。

补充有机质：对土壤补充有机质，提高土壤肥力，有利于土壤有益微生物的培养，使土壤保持疏松且不易板结，同时也可以调节土壤的酸碱度，同时又有促根、养根、护根的功效，促进植株肥水吸收更全面，为培养健壮枝条与结果母枝打好基础。

2）清洗树体

近些年来，芒果细菌性角斑病越发严重，甚至全园都会感染，造成枝条死亡，果面密布病斑，严重影响果面，影响果品商品价值。细菌性角斑病也是芒果主要病害之一，传染

速度非常快,需要通过清洗树体等措施进行防控。

在采果后修剪清园时,将修剪的枝条清理至园外集中烧毁,同时对残留的病菌进行消毒处理,对树干枝条及树冠周围喷施高尚 750 倍液或 50% 春雷·王铜可湿性粉剂 1 500 倍液 2~3 次进行防控,以清除细菌性角斑病、炭疽病、红点病等病菌,降低病原菌基数,防止造成大的经济损失。

3)促梢、萌梢

萌梢的整齐度、枝梢梢芽的饱满与健壮度直接关系枝梢的质量,进而影响结果母枝的培养、树体的成花与果实产量、品质。同时枝梢萌芽不整齐也给管理工作带来诸多不便。因此促梢是一个重要环节,这个关键点没有做好,就会影响后面一系列管理工作。要促好梢,培养健壮枝梢就必须养树,养树就必须改土、促根、补充有机质及植株所需的大中微量元素,要土壤与枝梢双管齐下,才能更好地培养健壮枝梢。

3.2.5 水分管理

3.2.5.1 灌溉

灌溉方法有沟灌、树盘浇灌、喷灌的滴灌等。通常在花芽分化前 60~90 d 应尽量保持干燥,植株在正常的花芽分化后会陆续抽穗→开花→结果→膨大,这段时期适逢芒果开花及果实迅速膨大期,也为最需水分的关键时期,至果实→硬核→成熟→收获,前后约有两个月的时间,为需水量较少时期,应保持土壤干燥,才能促使果实增加甜度及提高品质。

3.2.5.2 排水

芒果园忌积水。芒果较耐旱但不耐涝,这是因为芒果树根系呼吸作用强,需氧量高,如排水不良,将会抑制根系呼吸,降低吸收功能。同时,当土壤含水量高时,会抑制花芽分化,促进枝梢之生长。长期积水会导致落叶、落花、落果、枯枝、烂根,影响树体生长发育。

3.2.6 开花坐果管理及产期调节

3.2.6.1 开花坐果期管理

开花及坐果期是芒果生产管理上极为关键的一个时期,直接影响到芒果的产量和品质。因此,针对不同生长阶段制定不同的管理方法,是提高芒果产量与果实品质最有效的方法。开花及坐果期的管理有以下几个方面。

(1)开花枝梢的整理

在花芽分化前约 2 个月疏除过密枝、萌弱枝及病虫枝,每枝只留 1~2 个梢即可,这样可增加树冠的透光度,利于养分的集中和形成良好的通风光照条件,促进花芽分化及防止病虫害发生(图 3-8)。

图 3-8 芒果花穗

（2）授粉昆虫的饲养

芒果的授粉主要靠蝇类，可在芒果抽穗时饲养蝇类。芒果自抽穗至小花开放需 14~20 d，饲养蝇类自产卵→成虫的过程也需 14~20 d，所以在抽穗时饲养蝇类，待成虫后正逢小花开放，便可获得最佳的授粉效果。饲养方法：可在果园内放置猪内脏或臭鱼，引诱蝇类产卵，增加苍蝇的数量，促进芒果授粉坐果。

（3）微量元素及钙肥的施用

芒果自开花至幼果期对微量元素及钙的需求量较大，可在开花前 10 d 及盛花和谢花后喷 1 次 0.2% 硫酸锌和 0.2% 硼砂溶液，以促进授粉及结实。

（4）疏果及修剪果枝

在幼果如花生大小时应进行疏果，疏掉病虫果、畸形果，以提高果实品质和商品果率。一般每穗只保留 2~4 粒果，保留的位置以中央为佳。另外，此时期还要剪掉刚抽出的嫩梢。

（5）促进着色

有些果皮为红色的品种，如红芒 6 号、爱文等，在中果期修剪时应去除遮蔽果实的阴弱枝、病虫枝等，使果实接受充分光照，着色均匀。

（6）套袋

芒果袋的大小依品种而异，对金煌、象牙等果形较大的芒果品种，可采用长 30 cm、宽 20 cm 的纸袋。纸袋材料可由白色蜡纸、黑色牛皮纸或银色牛皮纸及无纺布组成。

套袋时间一般是在坐果基本稳定后，即第二次生理落果结束，果实生长发育到鸡蛋大小时为宜。太早套袋会导致后期空袋多，浪费人力物力，套袋太晚则失去效果。果皮为红色的芒果品种套袋时间与普通品种不同，如爱文在采收前 30~50 d 时进行套袋会更利于果实着色；金煌芒可提早套袋，套袋后果实表面会较细致光滑，且有良好的果粉产生；凯特为晚熟品种，若提早套袋会引起果实外表着色不良，因此可稍晚套袋以利果实着色（图3-9）。

图 3-9　芒果套袋管理

套袋方法：在果实套袋前喷杀菌和杀虫混合剂，也可将所用的袋放入配有杀菌杀虫剂的药液中浸湿后使用，选择发育正常、无病伤的果逐个套上，然后扎紧袋口和果柄，袋底留漏水孔，以排除袋中积水。

3.2.6.2 产期调节

（1）花期的调节

海南芒果的催花时间一般在8—10月进行，这一时期都处于秋梢期，常用硝酸钾和多效唑进行调控。多效唑可土施，也可以喷施，土施有效期比较长，3年内都有效果，但土施应提早进行，一般在第二蓬梢抽出10~15 cm长时就可土施，多效唑土施方法为在芒果树冠投影（滴水线）内挖深15 cm、宽10~15 cm的环沟或距树头40~50 cm处对称两侧各挖1条深15 cm、宽10~15 cm、长40~50 cm的直沟，将多效唑兑水均匀后施于沟内，覆土。每平方米树冠投影面积施8~10 g。喷施时用硝酸钾，与多效唑、爱多收、奈乙酸、乙烯利、控梢灵、杀梢灵、和矮壮素等1~2种混用。施用量应根据树冠大小、树势强弱确定。一般为健壮树每株施15%多效唑25~40 g，弱树适当减少。

（2）提高坐果率及保果

在提高芒果坐果率、药物措施保果增产方面，可在盛花期喷70 mg/L的赤霉素和30 mg/L的萘乙酸来提高坐果率，增产效果明显。一般常用0.3%硼砂或0.2%硼酸喷施，每隔10~15 d喷一次，上午露水干后喷，或者下午喷，注意避免在中午太阳暴晒时喷施。

（3）采用早、中、晚熟品种搭配

采用早、中、晚熟品种搭配，拉长市场供应期，提高市场竞争力。目前我国芒果产区品种种植较为单一，基本上都是中熟品种，成熟期较为集中，势必影响到果实的销售处理。因此对于区域性种植的芒果来说，推荐采用早、中、晚熟品种搭配种植，以达到产期的调节。

（4）利用产区差异调节市场

采用产期调节技术后，海南三亚2—4月就有早果上市，而广东湛江坐果最早也要到6月上中旬，四川攀枝花又要比广东迟30~40 d果实才成熟，因此可利用不同地区气候差异及不同品种成熟期的差异来进行市场调节。

4 芒果主要病虫害防治与采后保鲜技术

作为热带、亚热带主要水果，芒果产业近些年得到了快速发展。当前我国芒果种植面积已超过500万亩，虽然我国芒果单位面积产量有了大幅度提高，但品质却有所下滑，病虫害作为影响芒果品质的主要因素之一，发生日趋严重，已成为芒果产业发展的障碍，严重威胁到芒果生产的安全。

4.1 芒果主要虫害及其防治

4.1.1 芒果切叶象甲

4.1.1.1 为害症状

成虫取食嫩叶表皮和叶肉，啃食斑近乎圆形，仅余下透明状下表皮，叶片卷缩干枯。雌成虫在嫩叶上产卵，并从叶片近基部处咬断，切口齐整如刀切，造成秃梢，严重影响树势。

4.1.1.2 生活习性与发生特点

芒果切叶象甲年发生代数因地区而异，海南每年发生9代，广西每年发生7代，云南西双版纳地区每年发生3~4代。冬季无越冬现象。成虫产卵剪叶期间，若遇烈日，这些落地带卵剪叶迅速萎蔫，最终会导致卵和幼虫大量死亡。土壤含水量对蛹的生长发育影响较大，低于10%或高于20%均能引致前期蛹死亡。芒果切叶象甲（图4-1）成虫羽化出土后，有明显向上性、趋嫩性和群集性，常聚居于芒果嫩梢、嫩叶上。遇惊动时假死落地或飞走。

图 4-1　芒果切叶象甲

4.1.1.3　防治方法

农业防治。①拟新开的果园，要避免芒果和龙眼混栽，以杜绝或减少虫源；②对芒果与龙眼混种的果园，可结合除草、施肥或控制冬梢，进行翻松园土，杀死在土壤中的部分虫蛹和越冬幼虫。

人工防治。在此象甲发生期，要及时收集被咬断落地的芒果嫩叶，消灭虫卵，降低下一代的虫口。在芒果第二次生理落果开始至收获期，经常注意捡拾地面落果及遗弃的果核并烧毁。

化学防治。在各代成虫羽化期，掌握虫情，适时喷药杀死成虫。有效农药有：90%晶体敌百虫，或80%敌敌畏800~1 000倍液，或20%速灭杀丁，或2.5%敌杀死2 000~2 500倍液，或80%敌敌畏与40%乐果各1 000倍的混合液。

严格执行检疫制度，严防害虫随果实、果核或种苗向疫区外传播。疫区一经发现，必须及时扑灭。

4.1.2　脊胸天牛

4.1.2.1　为害症状

幼虫钻食芒果、腰果等果树的枝条，通常从幼枝条侵入，钻蛀枝条和树干，使枝条干枯，影响植株的生长势，刮大风时常造成断枝或树干倒折，最终导致全株死亡。

4.1.2.2　生活习性与发生特点

1年发生1代，幼虫孵化后即钻入枝条髓部蛀食，在隧道末端常有白色胶状物聚集。幼虫老熟后即在隧道内化蛹，蛹羽化后在蛹室内停留一段时间后外出活动，啃食幼嫩枝条、嫩芽、嫩梢皮（图4-2）。

图4-2 天牛幼虫与成虫

4.1.2.3 防治方法

加强果园管理。增强树势,提高树体抵抗力。结合修剪及时剪除受害枝条,清理果园,将病残物销毁,减少虫源。

人工防治。根据其生活习性可用铁丝捅入蛀口,进行人工捕捉。为害严重的果树,可进行截冠复壮。

化学防治。在被害树干中可用2.5%溴氰菊酯或20%氰戊菊酯填堵蛀口。

4.1.3 蓟马

4.1.3.1 为害症状

蓟马是芒果常见虫害,该害虫以成虫、若虫锉吸芒果的茎、叶、花、果的汁液,导致植株枯萎。被害叶片叶缘卷曲不能展开,呈波纹状,叶片变狭或纵卷皱缩,叶脉淡黄绿色,叶肉出现黄色挫伤点,似花叶状,最后叶片失去光泽、僵硬、变黄、变脆、脱落。新梢或幼苗顶芽受害,枝叶丛生现象或顶芽表现萎缩(图4-3)。受害的果面表皮油胞破裂,逐渐失水干缩,疤痕随果实膨大而扩展,呈现木栓化形状。

4.1.3.2 生活习性与发生特点

芒果的蓟马发生高峰和芒果物候期有密切的关系,蓟马为害从初花期开始,至盛花期为害种群数量达到最大,当小果期到来时,虫口数量会明显下降。在芒果生长、开花结果的时候,遇到温暖干旱的天气,为害会更明显。

4.1.3.3 防治方法

抓好花前修剪工作,疏除过密的枝叶,增加树冠的透光度。也可以在花期时每2~3 d摇树枝抖落残花,其目的是扰乱蓟马晒息场地,减少虫口密度。

可用10%吡虫啉可湿性粉剂1 500倍液和0.1%阿维菌素可湿性粉剂1 500倍液交替

防治，也可以将吡虫啉和阿维菌素两种药剂混合后按 1 500 倍液喷洒使用，在花蕾期与谢花后各喷 1~2 次。

图 4-3 蓟马及其为害

4.1.4 横纹尾夜蛾

4.1.4.1 为害症状

横纹尾夜蛾，俗称钻心虫，其幼虫往往钻进芒果的幼嫩树梢，蛀食芒果的花穗、嫩叶等部位，导致树梢枯萎，落叶，严重影响植株正常生长，导致结果量下降。

4.1.4.2 生活习性与发生特点

一般一年繁殖 8 代，世代重叠，成虫昼伏夜出，趋光、趋化性弱，幼虫一般 5—6 月和 8—10 月为害嫩梢，10—12 月和 2—3 月分别为害花蕾和嫩梢（图 4-4）。幼虫共 5 龄，老熟幼虫在芒果的枯烂木、枯枝、树皮或其他虫壳、天牛粪便等处吐丝封口化蛹。

图 4-4 尾夜蛾及其为害

4.1.4.3 防治方法

人工防治。在尾夜蛾集中发生期,可采用主干绑扎塑料包椰糠(木糠)诱集横纹尾夜蛾。剪取 10 cm 宽的农用塑料薄膜环绕树干一圈,并留 10 cm 长的接口重叠,下端用包装带紧紧绑在树干上,拉开薄膜上端使之成喇叭状,填入湿润椰糠(木糠),以诱导横纹尾夜蛾幼虫在其中化蛹。

生物防治。保护和增殖寄生天敌;养鸡灭虫。

化学防治。当芒果抽梢、花穗 3~4 cm 长时进行喷药,可保护嫩梢免遭此虫侵害。有效药剂有:40% 乐果乳油 1 000 倍液、50% 稻丰散乳油 800 倍液;在横纹尾夜蛾产卵期可分别用 90% 敌百虫,或 20% 杀虫畏各 1 000 倍液杀卵。

4.1.5 瘿蚊

4.1.5.1 为害症状

该虫以芒果嫩叶为食料,被害嫩叶先见白点后呈褐色斑,穿孔破裂,叶片卷曲,严重时叶片枯萎脱落以致梢枯(图 4-5)。

4.1.5.2 生活习性与发生特点

在温暖潮湿的季节大量发生和繁殖,成虫喜欢潮湿荫蔽的环境,怕强光照射。被幼虫

图 4-5 瘿蚊及其为害

叮咬吸汁后的叶片伤口呈黄白色瘤状，叶片皱缩。湿度较高时，特别是雨天过后，伤口易感染炭疽病，严重影响芒果新梢的生长。

4.1.5.3 防治方法

结合修剪及时清园，保持果园内树冠通风透光。及时除草，适时松土，以破坏瘿蚊滋生、繁殖及化蛹场所。同时注意合理施肥，科学用水，促进新梢抽发整齐健壮，提高抗逆性。在瘿蚊已发生地区，采果后清园时需要将病虫枝叶、内膛阴枝剪去，并集中园内枯枝落叶一起无害化处理。

新梢期是瘿蚊防治的关键时期。新梢嫩叶抽出 3~5 cm 时，喷洒 75% 环丙氨嗪可湿性粉剂 5 000 倍液或 25% 溴氰菊酯 2 500 倍液或 48% 氯吡硫磷 1 000~1 500 倍液，梢期防治 2~3 次。

由于瘿蚊幼虫每年会入土化蛹越冬，可在每年 2—3 月春雨前施用辛硫磷颗粒剂加细沙或泥粉，均匀撒在树冠之内的土表，然后覆土，以杀死土中幼虫及羽化出土成虫。

4.2 芒果主要病害及其防治

4.2.1 炭疽病

炭疽病是芒果生产中发生最普遍、危害性最大的一种真菌病害，可通过侵染芒果叶片、嫩梢、花和果实，造成生长期叶斑、梢枯、落叶、落花和果腐。该病在高湿情况下易大量发生，芒果花期和幼果期较易感病，成熟果易感病，嫩梢期发病最重（图 4-6）。

4.2.1.1 为害症状

该病为真菌性病害，一年四季均可发生，潮湿、多雨的季节尤为严重。病叶初期出现褐色小斑点，周围有黄晕。病斑扩大后呈圆形或不规则形，黑褐色，数个病斑融合后形成大斑，使叶片大部分枯死。果实感病后，只呈现针头般的小黑斑，此症状一直潜伏至果实成熟时才会暴发出来。而且其潜伏期即具有较强的感染性。

4.2.1.2 防治方法

采用田间卫生 + 田间管理 + 化学防治 + 采后处理的防治模式。

田间卫生：修剪、清除、烧毁、选种抗病品种（台农一号等）。

田间管理：通风透光、清除杂草。

化学防治：可以根据芒果的生长期及天气情况来及时喷药防治。比较常用的杀菌剂有 1∶100 的波尔多液 500 倍液，40% 多菌灵 200 倍液，25% 代森锌 400 倍液，75% 百菌清 500 倍液，70% 甲基硫菌灵 1 000~1 500 倍液等交替使用均有防治效果。此外，用 1.5% 多

图 4-6 芒果炭疽病

氧霉素 300 倍液，在花蕾期每 10 d 喷 1 次，连续 2~3 次。果期可以每月喷 1 次，抽梢期自芽萌动伸长时起，每 7~10 d 喷 1 次，连续 2~3 次。

采后处理：果实成熟后及时采收，尽量减少伤口，果实采后用清水浸泡清洗。

4.2.2 白粉病

4.2.2.1 为害症状

为真菌性病害，集中发生在气候比较干燥的地方。主要侵害芒果的花穗、叶和幼果，花期至幼果期是其主要发生期，可引起严重的落花落果，进而影响坐果率。比较明显的特征是病部密被一层白色粉状物，感染初期只呈现白色粉状的小斑点，而后逐步扩大为大片的斑块，形成一层白色的粉状物体。当芒果花被感染后，其花萼、花茎和枝柄也逐渐被这种白色粉末所覆盖，而后变为黑色，直至枯萎（图 4-7）；当芒果幼果被感染后，果实表面便会布满白色粉末，导致果实极易脱落。

4.2.2.2 防治方法

物理防治：做好田间卫生，集中清除或者烧毁枯枝败叶等。

化学防治：常用药剂如硫黄制剂、粉锈宁、灭病威、硝螨普等。

喷药时机：重点时期是花期。抽蕾花序 5~10 cm 后开始喷药，每 10~15 d 喷施 1 次。

或在抽蕾、盛花和稔实期用 320 筛目硫黄粉喷粉 1 次。盛花期降低浓度，避免药害。

农业防治：控制过量施用化学氮肥，增施优质有机肥和磷钾肥。培育具有良好抗性的芒果品种，如秋芒、粤西 1 号、吕宋芒等。

图 4-7　芒果白粉病

4.2.3　霜霉病

4.2.3.1　为害症状

霜霉病由低等真菌引起，在潮湿天气及环境下易发生，通过气流传播，主要发生在芒果花期。花期感染霜霉病，花序会产生白色或紫灰色、灰褐色、黑色等颜色霉层，造成黑花、烂花、不挂果，轻者减产，重者绝收。一般在芒果花特别多且都挤在一起，透光性和通风性不好及芒果出现团花的时候容易感染霜霉病（图 4-8）。

4.2.3.2　防治方法

一是加强栽培管理，冬季彻底清园；二是完全谢花后，及时进行摇花、洗花处理；三是主花序 3~5 cm（花序生长 5~7 d）时，进行拉花处理，避免团花；四是人工对花序进行疏花，短截花序（部分晚熟区域），增强通风透光性

图 4-8　芒果霜霉病

能；五是在初花期及谢花初期通过药剂进行化学防治。

4.2.4 蒂腐病

4.2.4.1 为害症状

该病主要为害芒果果实，初期病部（果蒂附近）生水浸状褐色不规则形的病斑，而后呈暗褐色至黑色，并迅速向下发展。初时蒂部暗褐色、无光泽，病健部交界明显，在湿热条件下，病部向果身扩展，病果皮由暗褐色变为深褐色或紫黑色，同时，同肉组织软化、流汁，有蜜甜味，3~5 d 全果腐烂变黑，病果皮出现密集的黑色小粒，此为病菌的分生孢子器（图 4-9）。孢子角黑色、有光泽。从果皮伤口或皮孔侵入，引起皮斑。该病也可以为害枝条，引起裂皮、流胶。从枝条剪口侵入，引起剪口"回枯"。

图 4-9　芒果蒂腐病

4.2.4.2 发病特点

初侵染源来自枯枝、树皮和落叶；传播途径主要是雨水，成熟分生孢子在无菌水中 4~5 h 后萌发，分生孢子随雨水从受伤的果柄、果实剪口或机械伤口侵入，果实成熟前病菌处于潜伏侵染状态，果实后熟以后迅速腐烂。

一般来说，此病是前期发生侵染、潜伏，接近成熟至采收后再出现病症，所以生长期的防控及采收处理都应重视。前期用药可参考炭疽病，采收时一果两剪，减少机械伤，并用咪鲜胺等浸果处理。

4.2.4.3 防治方法

搞好果园卫生，减少初侵染源。果园修剪后应及时处理枯枝烂叶，修剪时应尽量贴近枝条分枝处剪下，避免出现枝条回枯。果实采收时采用"一果二剪"法，可降低病原菌从果柄侵入的速度和概率。所谓"一果二剪"，即在果园采收时的第一次剪，留果柄长约

5 cm，采后到加工厂处理前，进行第二次剪，留果柄长约 0.5 cm。放置时果实蒂部朝下，以防止胶乳污染果面，每剪 1 次，都需用 75% 酒精蘸果剪。采果前不要施用含钙化合物，如含钙叶面肥等。将采收处理后的果实置于 10~13 ℃的温度下储藏，也可以延缓蒂腐病的发生。

4.2.5 流胶病

4.2.5.1 为害症状

芒果种植过程中的真菌性病害之一，主要为害芒果的树干、枝梢和叶柄，引发枯干，也可为害芒果的幼苗以及果实（图 4-10），在我国多地主要的芒果种植区均有发生。流胶的本质是因为树胶多糖合成过多所致，高温高湿及荫蔽的环境都会导致流胶病的发生。

4.2.5.2 防治方法

一是加强果园的栽培管理。增强树势，合理浇水，增施有机肥，以磷钾肥为主，提高植株抗性，培养健壮树势，缓解环境对树体的胁迫。

图 4-10　芒果流胶病

二是及时清理果园。结合整形修剪，及时剪除染病枝梢（病部以下 20~30 cm 处剪除），集中带离园外烧毁，减少果园初侵染源，并对伤口进行消毒处理，可以采用波尔多液进行清园消毒（注意喷施到叶背处和内部枝梢）。

三是加强对天牛、横纹尾夜蛾等害虫的防治。要注意防治病虫害，减少伤口的产生，可以采取对树体涂白等方式，此外，也要注意在栽培过程中减少伤口产生。

四是主干上的病部，可以用消毒的刀割除病部，割至健康组织处，再用氢氧化铜或硫酸铜钙或苯醚甲环唑涂抹伤口。

4.2.6 露水斑

4.2.6.1 为害症状

露水斑是芒果生长期间出现的一种病害，病原主要是球孢枝孢菌、枝状枝孢霉。感染该病的芒果果面呈现似"露水"状的病斑，且在早晨露水凝聚于果面时病斑比较显眼，果农俗称"芒果露水斑病"，对芒果外观品质影响极大，其发病与果皮厚度、韧性及树势等有关，且越接近成熟的果发病越重（图 4-11）。

4.2.6.2 发病特点

果面营养丰富的芒果品种上更易发生该病;密植遮阴、枝叶茂密、不通风透气的果园,特别是管理粗放、修剪不到位以及山区或低洼处高温高湿且杂草生长旺盛的老果园发病较重;有研究结果显示,温度28 ℃且相对湿度大于90%的条件适于露水斑病原菌的侵染,故持续雨雾天气和久逢"头水"等特殊天气情况,土壤及空气的湿度较大,特别是5—6月遇多雨高温高湿天气状况下发生严重;多次使用有机乳油类农药,尤其是经常使用噻苯隆、赤霉素、乙烯利等的果园易受侵袭,发病率较高。该病初发现一般3 d左右就会出现大量病斑,树冠内膛枝果实上首先出现

图4-11 芒果露水斑

症状,继而从树冠下面枝条的果实逐渐往顶部果实传染,同时发现相同枝条上的两个果中,如果一个果实感病,另外一个果实很快也会产生同一症状。

4.2.6.3 防治方法

一是选用抗病品种。结合芒果种植区域内的气候特点,因地制宜选择抗病品种,栽种无病苗木。不同物候期的品种不要混种,尽量做到不同的芒果品种按片按区域规划种植。

二是加强修剪,适时整枝。芒果采摘后至新梢抽生之前应利用晴天对树体进行合理修剪,剪口应及时涂抹愈伤防腐膜,以助于其伤口的愈合,防止病菌的侵入。在修剪的同时进行整枝工作,打掉闲枝,剪除过密枝、交叉枝和病虫枝等,及时清理荫蔽的芒果枝条,控制树冠,同时进行疏花疏果,以利于保留果树适宜的挂果量,增强果园以及树冠的通风透光性。

三是清洁田园、及时防治媒介昆虫。做好清园工作,在芒果花期前彻底清理病枝和枯枝等并将其集中烧毁,或深埋于土中,或集中堆放后喷布速腐剂加速其腐烂,减少初侵染菌量。选择非病原菌寄主的低矮草种在芒果树行间进行生草覆盖,同时注意定期刈割,保持行间草种的适宜高度和生长量,以利于保持果园的生物多样性和生态平衡。

四是及时套袋护果,降低病害侵染果实的概率。一般在第二次生理落果后随即对芒果进行套袋处理,及时保护果实,避免其遭受病害的侵染。套袋处理时应保证袋口的折叠要利于导流,捆扎应严密,防止雨水进入袋内。此外,应及时锯除已枯死的大枝条,同时用波尔多浆封闭保护锯口,减少病害传播的概率。

五是合理水肥、增强树势。加强水肥管理，力争做到配方施肥，均衡树体营养，提高其抗病性能，促使果树抽梢和开花整齐有序，以便于统一进行田间管理。首先，做好果园的排灌设施，注意雨天及时排水，避免积水增加果园的湿度。其次，增施有机肥，一般可在芒果幼果期结合修枝，将有机肥和有益菌肥挖沟埋入树冠根部，促进树体健壮，以提高其抗逆性。最后，注意营养元素尤其是钙、硼、镁、钾等的合理应用，可在芒果扬花期以及小果到膨大期叶面补充钙硼肥1次，以利于芒果上粉，减少水滴直接接触果皮的概率；在膨大至着色期可每10~20 d喷施寡糖素1次，同时补充钾和镁，增加芒果长势，增加果皮内层厚度；此外，应避免在芒果上使用含隐性添加违规成分的叶面肥。

六是合理使用植物生长调节剂，减少强内吸性有机乳油类药剂的持续使用。在芒果生长早期可使用生长调节剂进行统一催花以利于开花整齐，但应注意生长调节剂的使用剂量和次数，避免花穗经久不谢以及谢花后聚集成簇不掉继而滋生出大量的露水斑病菌；在芒果生长的中后期应尽量减少或间隔使用噻苯隆、乙烯利等生长调节剂，避免其损伤果皮以及破坏果粉，减少露水斑的形成条件。

七是科学使用化学农药防治露水斑。用化学药剂防治芒果露水斑时应尽量减少强渗透性农药的使用量和次数，避免因药剂导致果皮通透性的增加，具体要求：第一，芒果园修剪后以及芒果开花前喷施石硫合剂或波尔多液预防露水斑发生。第二，可在芒果挂果初期施用多菌灵、百菌清等内吸性杀菌剂进行保护。第三，在果实发育后期做好监测和防治工作。注意在大雾、重露和雨后等特殊天气要强化监测，观察果实受害情况，当田间发病率达3%时立刻施用药剂进行防治。

4.3 芒果采收及保鲜技术

4.3.1 采收

4.3.1.1 确定采收成熟度的方法

（1）外观鉴定

①果实已停止长大，达到该品种的一般重量。②果皮颜色发生变化，果皮由青绿变为苹果绿、黄绿或淡绿，有些品种果皮出现白蜡层或皮孔微裂。③树上发现有成熟果，或有果实蝇蛾为害果实。④果肩形态变化，成熟果实果肩发育饱满、浑圆，有该品种的特性。⑤果肉硬度，用果实硬度计测果肉硬度，一般为0.172~0.196 MPa。⑥果肉颜色，由乳白到白黄到淡黄则达到成熟度。剪下果实以流出的汁液乳白黏稠即达采收成熟度。

（2）果实的生长日数

一般从盛花期开始计算，不同成熟期的品种各自发育日数不同，如早熟品种粤西1号的生长日数为85~100 d，紫花芒、桂花芒生长日数为109~125 d等。高温干旱可提早成熟，湿度大或雨水多会推迟成熟。

（3）内含物的测定

pH值一般小于3.2，可溶性固形物一般在4%~6%，不同品种差异较大。如吕宋芒总固形物达6.5%，柠檬酸最高值为2.5%时达到青熟标准。

另外，还可采用比重法，即达到成熟的芒果比重大于1，投入水中下沉或半下沉者达到采收成熟度。

4.3.1.2 采收技术

采收过程中应防止一切机械损伤，如碰伤、擦伤、摔伤等，还要尽量避免果柄流胶污染，伤害果皮，要严格按照技术要求操作，切实做到无伤采果。采收时注意以下几点。

一是采收时间宜选在9:00露水干后及16:00—18:00不受太阳暴晒时进行。此时果柄排胶较少，以减少污染果面。此外在雨天不宜采果，以防病菌感染。

二是采果时宜采用"一果二剪"的方法。即从树上用果剪将整穗果或单果剪下，放入盛果的塑料篮中，运回室内后再用果剪在果柄0.5 cm长处剪断，以避免流胶污染果面。

三是收果后果实轻拿轻放，不能堆放在太阳下暴晒，也不能直接堆放在泥地和水泥地上，做到果实采收后不沾地。

4.3.1.3 果实的处理、分级及包装

（1）果实处理

果实在采下后8 h内应用清水或1%醋酸溶液洗去果皮表面的乳汁、泥污等，也可结合热药浴处理（见4.3.2）。

（2）分级

应按商品标准要求分级。一级果为果形端正，形状大小一致，表皮光滑，无病斑和害虫叮咬痕迹及其他损伤；二级果为无严重损伤（仅少数果有轻微疤痕）。凡是受果蝇或吸果夜蛾、蒂腐病为害，或炭疽病斑多的果实均不能作商品果。另外，也可按品种类型的果实大小分级，在同一箱中品种相同、大小一致。

（3）包装

经过洗涤或热水及热浴处理过的果经过晾干、冷却，用洁净、柔软白纸或厚0.11 cm聚乙烯薄膜单果包装，然后把果蒂向下，果实凸凹面向上摆放，以防止果实包装后流出汁液损害外观。放入能容纳不同重量的瓦楞纸板箱内。

4.3.2 储藏保鲜

4.3.2.1 采后的保鲜处理

（1）防腐处理

果实采后热水浸果是许多芒果商业包装必须采取的措施之一，55 ℃热水浸果 5 min，50~55 ℃热水浸果 15 min，47 ℃热水浸果 20 min 能有效地控制储运期间炭疽病的为害，在热水中加入苯菌灵、涕必灵之类的杀菌剂浸果，效果更佳。

（2）涂膜

涂膜能降低果实呼吸代谢和储运过程失水失重，还能抑制微生物侵害与生长，可用 6% 脱蜡紫胶液（含 0.25% 联苯）涂膜芒果。

（3）化学处理

化学处理是芒果处理中延长寿命的一种方法，使用该法时可利用激素延长果实的后熟时间，利用浸果增加果实硬度，使绿熟果实延长储藏时间。用 0.1% 的表面活性剂梯普（阳离子洗涤液）、浓度为 100~200 mg/L 的 GA 液处理在 28~37 ℃储藏明显延缓果实成熟，失重较少。另外，用 32 mg 的乙烯化氧处理后，经 16 d 室温储藏，果皮呈金黄色，风味好，此外，1−甲基环丙烯、钙等化学药剂对芒果的处理均能在降低果实储藏期的生理失重、延缓后熟、保持果实硬度等方面有一定作用。

4.3.2.2 包装

一般用洁净、柔软的白纸或厚 0.1~0.2 mm 的聚乙烯薄膜进行单果包装作为内包装，然后放入能容纳 5 kg、10 kg 或 20 kg 的瓦楞纸箱内，纸箱内分两层；每层之间用纸板隔开，每层又分 20~30 个格，每格放 1 个芒果，格子大小应与果实大小相合，然后在果实上贴上色彩明显、设计简要的小商标，起到美化商品和宣传的作用。

4.3.2.3 储藏

（1）低温储藏

作较长期储藏或运送的芒果，应采用低温储藏，适当的低温能延长芒果的储藏寿命，芒果最适储藏温度 9~13 ℃，相对湿度 85%~90%，但各品种间或处理方法不同都有差异。

（2）气调储藏

用气调储藏来调节芒果储藏期间氧和二氧化碳浓度的比例，推迟呼吸高峰的出现是有效果的，气调机有燃烧式、裂解氨式和分子筛吸附式等类型。其中以焦炭分子筛气调机较为实用，有降低氧气、二氧化碳和乙烯的多种功能。

（3）常温储藏

常温储藏一般作为短途运输和储藏时间较短时的储藏方式。用 500 mg/kg 多菌灵热水浸果后晾干、套袋，在常温下（30 ℃左右）储藏保鲜期可达 10~15 d，一般来说，常温保存期都不能超过15 d，否则品质下降，风味变淡。在常温储藏时一定要注意包装箱留取筐空隙，以利通风散热。

4.3.2.4 催熟

商业上一般采用人工催熟来促进后熟，用500~2 000 mg/kg乙烯利直接浸泡果实5 min，可使芒果成熟天数为4~6 d。另外，乙烯利和脱落酸处理加速芒果后熟，并可导致可溶性糖含量上升，从而改善品质，另外电石（碳化铝）用量为果重的1/2 000~1/1000，将芒果放入纸箱内，尽量不漏气，在果实的最下方放置用纸或布袋包住的电石，密封48 h后，电石吸湿后产生乙烯气体，达到催熟芒果的作用。

4.3.3 芒果加工产品

4.3.3.1 芒果原浆

芒果的收获期短，不易久存，在盛产地往往将大量剩余的成熟芒果尽快地加工成芒果原浆保存起来，为日后加工芒果饮料、芒果酱、芒果酒及其他芒果产品提供所需原料。芒果原浆的快速加工是高质量保存芒果最有效的途径之一，也是目前世界上通用的水果快速保存方法之一。传统芒果原浆的生产流程为：成熟芒果→清洗→选果→打浆→护色→胶磨→过滤→调配→脱气→杀菌→包装→冷却→成品储存，芒果原浆可把芒果的保鲜期从现在的不到50 d延长至1年以上（图4–12）。

4.3.3.2 芒果干

芒果加工中最常用的产品是芒果干，主要加工流程包括原料选择→清洗→去皮切片→护色处理→干燥→回软、包装（图4–13）。

（1）原料选择

选新鲜饱满，无腐烂、病虫害和机械伤的果实。选用干物质含量高、肉质厚嫩、纤维少、核小而扁薄、色泽鲜黄、风味好的品种。成熟度以八九成为宜，成熟度过低则芒果的色泽和风味较差，过熟易腐烂。

（2）清洗

将芒果倒入流动清水槽中逐个清洗干净，进一步剔除不合格果实，

图4–12 芒果原浆

图 4–13　芒果干制备过程与成品

最后按大小分级装在塑料筐内，沥干水分。

（3）去皮切片

用不锈钢刀人工削去外皮，修除斑疤，要求表面修削得光滑，无明显菱角。外皮必须去净，因果皮中含有较多的单宁，如未削净，在加工过程中容易产生褐变，影响成品色泽。去皮后的果实用锋利刀片纵向切片，厚度为 8~10 mm。残留果肉的果核可送去打浆制汁。芒果干的工艺流程及加工技术。

（4）护色处理

切好的芒果片果肉要立刻放入加有护色液，可达到保色、保味、防腐的目的。

（5）干燥

将护色处理后的原料均匀放于竹筛（浸硫处理的要先沥干水分），放入烘干机干燥。干燥初期温度控制在 70~75 ℃，后期控制在 60~65 ℃。干燥过程注意换筛、翻转、回湿等操作。

（6）回软、包装

待芒果干达到干燥要求的水分含量时，一般为 15%~18%，将产品置于密闭容器中，让其回软，时间 2~3 d，使各部分含水量均衡，质地柔软，方便包装。

4.3.3.3　新型保健饮料

芒果饮料是芒果主要加工产品，目前主要有芒果乳饮料、复合饮料、果肉饮料等（图4–14）。

图 4-14 芒果饮料

4.3.3.4 芒果全粉

芒果全粉是将新鲜芒果经真空冷冻干燥精制而成的天然果粉,其组织结构细腻,易溶解,保持了芒果原有的风味和营养成分,具有广阔的市场前景(图 4-15)。

图 4-15 芒果粉

4.3.3.5 芒果果酱

芒果果酱是一种老幼皆宜的食品,其风味独特,具有良好外观,保存性好,不仅可直接用于涂抹馒头、面包来食用,还可作为制作糕点的馅料,具有很强的市场竞争力(图4-16)。

图4-16 芒果果酱与芒果奶昔等

5 我国芒果产业现状与发展展望

5.1 我国芒果产业现状

在100多个生产芒果的国家和地区中，我国是世界第二大芒果生产国，品种资源丰富，种植区域广泛。2020年，全国芒果种植面积524.1万亩（34.94万 hm^2），总产量330.6万 t，产量约占全球的总产量的8.75%，产值达205.2亿元。同时，我国不仅是芒果的进口大国，也是芒果的出口大国，2020年芒果出口顺差为942.3万美元。

近年来，我国在芒果科研方面取得多项进展。跨区域、跨学科、跨部门、覆盖全产业链的芒果科技联合协作攻关体系初步形成，农业农村部芒果种质资源圃保存资源超过400份，并在国际上率先完成了芒果全基因组测序，选育出了金煌、贵妃、红玉、桂七及热品系列等一批优良新品种，构建起了早、中、晚熟优势区域布局体系，实现了芒果鲜果的周年供应。

5.2 我国芒果产业发展情况

5.2.1 近10年来我国芒果发展情况

截至2020年，我国拥有芒果园面积达34.94万 hm^2，芒果产业规模的快速扩张表现为产量、产值及种植面积的迅速发展，从芒果园面积来看，2011—2020 年，我国的芒果种植面积从 14.02 万 hm² 逐年增长至 2020 年的 34.94 万 hm²，增量达 20.92 万 hm²，增幅 149.26%，年均复合增长率约 10.68%（图 5-1）。

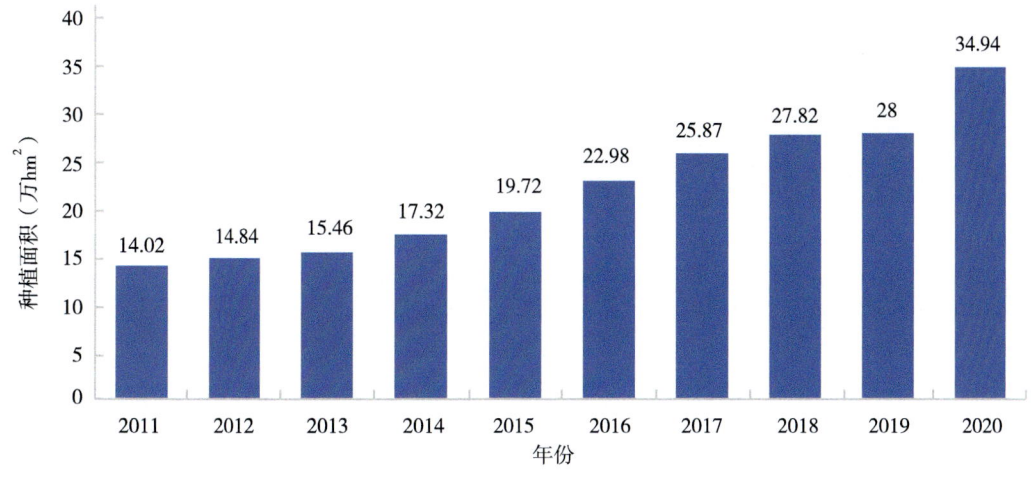

图 5-1　2011—2020 年我国芒果种植面积情况

近 10 年间，我国芒果单位面积产量从 2011 年的 7.16 t/hm² 波动增长至 2020 年的 9.46 t/hm²，增量达 2.30 t/hm²，年度复合增长率约为 3.14%（图 5-2）。

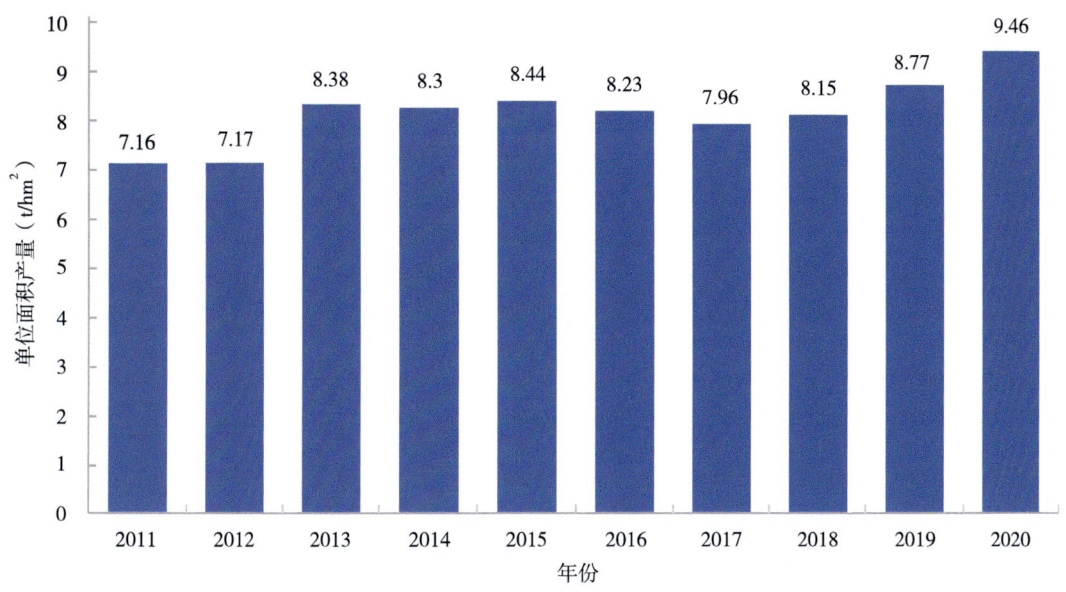

图 5-2　2011—2020 年我国芒果单位面积产量情况

5.2.2 我国芒果贸易情况

目前，在我国市场上获得芒果准入资格的国家主要有澳大利亚、巴基斯坦、秘鲁、越南、缅甸、菲律宾、泰国、厄瓜多尔、印度等，随着准入政策的变化，其他芒果生产国如果获得我国市场的准入资格，我国芒果市场的竞争将会更加激烈，国内芒果将受到进口芒果更大的冲击。

随着中国与东盟间的经贸往来密切，芒果作为东盟各国的优势产品，近几年大量出口中国，对我国芒果产业造成很大的冲击（图5-3）。

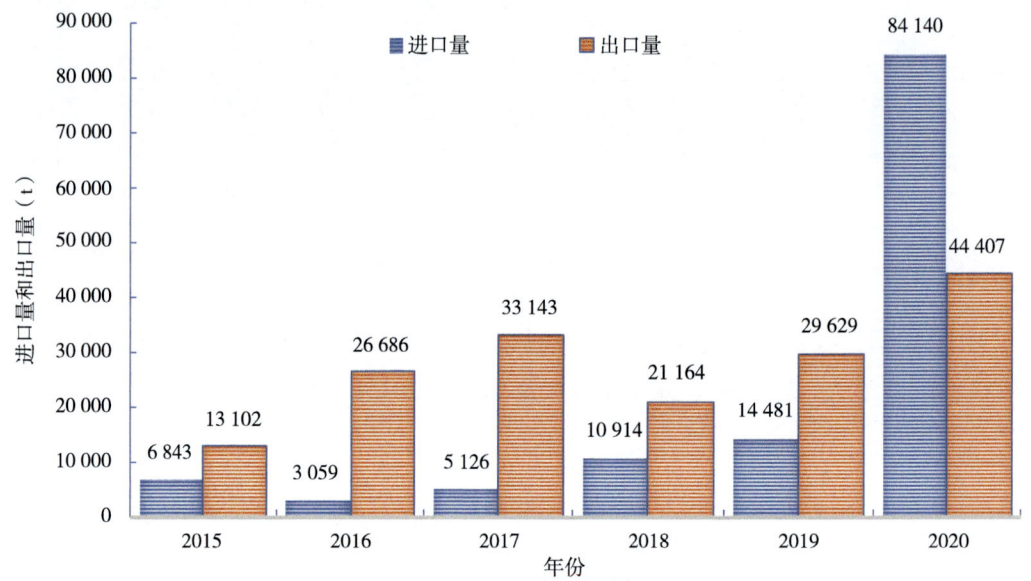

图5-3　2015—2020年我国芒果进出口情况统计
（资料来源：海关总署）

从进口结构来看我国进口芒果主要来自越南与泰国，其中越南进口芒果占到进口总量的79.84%。越南芒果供应期长且供应量充足，从每年的9月开始至翌年的6月结束，而且相比国内芒果而言有价格优势，从广东、广西等部分地区零售市场来看，越南芒果品种主要是大青芒，价格相对海南芒果略低，因此越南芒果拥有较强的竞争力。此外，来自澳大利亚的芒果卖相比较好，价格较高，针对中档市场；来自泰国的芒果如彩虹芒等定价较高，主要针对高端消费市场。

海关数据显示，2020年，我国芒果出口量达44 407 t，出口金额8 570万美元，主要出口至越南、俄罗斯、马来西亚等国，与2011年出口量3 867 t，出口金额282万美元相比，出口量增量达40 540 t，实现10.5倍的增幅，年均复合增长率约31.16%，出口金额增量达8 288万美元，增幅达2011年的29.4倍，年均复合增长率达46.14%。2020年随着

RCPE 协议的签订加上新冠肺炎疫情的影响，我国芒果进口量大增至 84 140 t。

2016—2020 年，我国出口的芒果产品几乎全是鲜、干芒果等初级产品，同时也进口大量鲜、干芒果。值得注意的是，该品类的贸易顺差总体呈下降趋势，2020 年跌破千万美元，降至 942.3 万美元。加工产品以芒果汁为主，虽然出口占比较小，但出口额从 6.5 万美元增长到 16.6 万美元，年均增长 26.2%，是鲜、干芒果年均出口增速的 2 倍多。随着中国企业加工技术和能力的提高，芒果加工品的出口比重有望进一步提高。

同时，我国也从泰国、越南、澳大利亚、秘鲁等地进口芒果，2016—2020 年，我国自越南、泰国进口均呈增长趋势，其中自越南进口增速尤为显著，自秘鲁、澳大利亚进口略有下降。越南既是中国芒果最大的出口目的地，也是最大的进口国。2020 年中国从越南进口芒果 6.72 万 t，占全国总进口数量的 79.9%；进口金额为 4 864.03 万美元，占全国总进口金额的 63.8%。值得一提的是，柬埔寨芒果获得中国市场准入后，2020 年 5 月首批柬埔寨芒果抵达中国市场，4 个月进口量达到 211 t。

与鲜芒果进出口相反的是，中国芒果汁的进口量远大于出口量。中国芒果加工品行业有着巨大的空缺，2019 年中国芒果汁的进口量为 1 805.7 t，出口量为 221.5 t，进口量为出口量的 8 倍。

我国芒果种植主要分布在海南、广东、广西、福建、云南、四川、贵州等多个省区。2020 年中国芒果种植面积 34.9 万 hm^2，总产量 330.6 万 t，产值达 205.2 亿元。根据数据，海南及广西的芒果产量占全国芒果产量的 50% 以上。云南、福建是芒果出口量最大的省份，2020 年云南省出口芒果 3.3 万 t，占全国总出口量的 73.9%；出口金额为 6 421.5 万美元，占全国总出口金额的 74.9%。

5.3 我国芒果产业发展存在的问题和风险

5.3.1 我国芒果产业发展存在的问题

一是劳动力成本不断攀升。我国芒果总体属于劳动密集型和技术密集型工作，包括生产过程中的修剪、打药、种植、施肥、采果、花果期等管理过程，劳动强度大，目前在生产中企业劳动力占比在 50% 左右，企业生产中的劳动成本不断攀升。

二是产业增速过快，区域性竞争和销售压力加剧。近些年部分产区面积增长超过 70%，产能迅速增长，在国内水果产能过剩和南美、东南亚水果进口的双重夹击下，如缺乏有效的销售渠道，丰产不丰收，价低难卖有可能成为部分地区芒果销售的常态。

三是组织化程度低，产业链条上各环节同质化竞争严重，导致产业整体竞争力下降。

目前我国芒果经营还以小农户为主，难以统一生产技术和生产标准，生产过程不规范，生产的果品质量参差不齐，产业化龙头企业数量少，采后储存、保鲜和运输技术不足，深加工和采后处理技术粗放，影响经济效益。同时，大量果农既是生产者，又承担技术员和销售员的角色，产业分工不明确，严重束缚着产业的健康发展。

四是芒果上市季节差异化红利优势削弱。我国是少有的利用地域特点和技术优势实现芒果鲜果周年上市的国家，但随着科技的发展，东盟、南亚大量国家也能实现芒果的产期调节，我国芒果产区差异化优势弱化，利润下滑。

五是信息化程度低，基础设施建设薄弱。信息流通不畅，市场信息不足，农民不能及时对产品结构进行调整，无法迎合市场需要，不利于市场发展。果农对果园资金投入不足，新技术、新产品得不到广泛推广应用，低产果园面积大，经济效益低。

六是创新不足，投入少。芒果科研呈现头重脚轻局面，重产前与产中，轻产后，标准化水平不足，产业创新研究与扶持投入少。目前我国芒果市场准入制度尚未建立，市场调节作用缺失，离标准化生产体系尚远。

5.3.2 我国芒果产业面临的风险

一是激素过量施用，通过催花生产早果现象大量存在。由于前些年不少芒果种植户催花采收早果，在春节前后国内芒果生产空档期卖高价，种植效益较好，形成了一定的带动效应，但在叶面或小果上喷施三无激素甚至直接用激素泡果，违反自然生长规律，引起芒果品质下降、食品安全、病害等诸多问题。

二是芒果产业链较短，深加工环节缺位。目前国内仅有10多家大型热带水果加工厂，深加工产能明显不足，每年芒果收获的季节除了果质较好易销的部分，剩下30%~40%的次果需由加工厂收购，价格低至0.4元/kg也难销售。芒果深加工环节的缺乏，使得芒果产业链得不到延伸，芒果附加值得不到提升，资源得不到合理配置，极大限制了芒果产业发挥应有的效益。

三是采后保鲜技术还未得到广泛应用。目前我国大部分产区的芒果基本不做采后保鲜，通常都是在产地采摘、分级、打包，收购商直接发往省外的批发市场。前些年，芒果多数在七成熟的时候就采摘了，除了为了抢早上市，也是为了在批发市场能保存久一些，因为果实硬度高很耐运。现在政府鼓励果农们不要摘嫩果，摘嫩果不但影响芒果口感，造成芒果后熟困难，甚至无法食用。芒果的采后热处理保鲜技术在科研领域已经较为成熟，但是因为农民对技术细节不了解，为了节省成本，很少使用。今后应加强农民对采后保鲜的认识，通过农民培训，提高技术水平，让芒果科学地保存和运输。

5.4 我国芒果市场与产业前景分析

我国的芒果市场潜力巨大，但需拓宽国内流通与营销渠道。近年来广西、云南和四川新种芒果面积较多，即将在3~5年后摘果，届时国内的芒果产量将迅速增长，同时柬埔寨走出去企业的新种芒果也开始收获直接销往国内，如果与潜在收获量匹配的流通与营销渠道欠缺，将出现鲜果供过于求，价格下跌的风险。

我国芒果价格高点多出现在每年12月至翌年2月，低点多出现在每年5—9月，主要是冬季新鲜水果供应较少，加上节假日期间（春节）消费旺盛而拉高芒果批发价格；而5—9月有大量新鲜水果上市，尤其是荔枝龙眼大量上市，消费者选择多，拉低芒果批发价格。

同时，运输环节的损耗和品质劣变也是影响芒果价值链的重要因子。现在只有少数企业意识到需采用现代化的物流方式减少运输环节的损耗，比如三亚树上熟芒果，产品单价高，但是皮肉薄，保存期短，不耐运输。2017年三亚某水果专业合作社借助全程冷链物流，通过生鲜电商平台向岛外卖出了超过100万kg树上熟的贵妃芒果。合作社专门购买了田间小型冷链运输车，芒果采摘后立马搬上冷链车，拉到空调制冷的恒温包装车间，打包好后直接由顺丰冷链物流运到销地市场的冷库或消费者手中。全程冷链物流除了增加冷链运输成本外，还额外增加了一次搬运成本，虽然费工、费时、费钱，但只有做好冷链运输，才能提高种植收益。

生产消费预期。国内消费未饱和，高品质芒果需求旺盛。我国芒果消费传统以鲜食为主，近年来芒果汁、芒果干、芒果酱等市场份额增长迅速，芒果价格向好，需求旺盛，鼓励了更多农户种植芒果。出于加工成本等因素的考虑，加工厂部分原料需从国外进口鲜果或者原浆来补充。未来若不出现异常气候等，芒果生产和消费将呈现良好的双轨发展态势。

市场前景分析。受国内消费的驱动，芒果单品在国内各地批发和零售市场都表现不俗。进口优质芒果主要在北上广深等一线城市的大型连锁超市销售，来自澳大利亚、泰国、秘鲁、厄瓜多尔等地，价格在每千克60元左右，远高于国产芒果市场价。澳大利亚芒果是高端超市里最贵的芒果，甚至达每千克200元左右，表明中国市场对优质芒果有较大市场空间。

5.5 展望

随着经济的发展和消费水平的提高,人们对芒果的消费需求也在不断增长,芒果生产已经成为我国热带亚热带农村经济发展中的支柱产业。但随着科学技术的进步以及生产经营环境的改变,传统的芒果种植受到强烈冲击,以品质为核心,节本、环保、安全、高效的现代化果树栽培管理与采后保鲜、加工技术成为新时代推动芒果产业健康持续发展的关键。目前我国芒果产业仍存在品种结构优化缓慢、产业链条较短、产品附加值较低、品牌效应整体不强等问题,下一步要着力打好芒果产业牌、市场牌和文化牌,通过科技更好地助力我国芒果产业的可持续发展。

一是优化芒果品种,提升芒果品质。要以科学规划为引领,统筹产业布局,通过整合优势科技资源,强化芒果种质资源圃和创新基地建设,加快发掘一批优异种质,选育一批高产优质突破性品种,建设良种繁育基地,提升芒果供种保障能力。

二是强化产业发展基础,延长产业链条。要集聚土地、资本、人才、信息等要素,以基地建设为依托,提高产业组织化水平,打造芒果"研发+生产+加工+销售+观光"的全产业链发展模式;加快推进果园、仓储、物流等基础设施建设,确保分拣处理和保鲜储运能力,实现由地头到餐桌的高效、低损流通;要加大芒果的精深加工,提高产品附加值,提升产业的风险防御能力。

三是拓展销售渠道,打造优势品牌。健全网络销售体系,拓展与国际采购商合作,培育芒果出口示范基地和出口企业。加快优势品牌培育,力争培育出一批国内领先、国际知名的区域公用品牌、企业品牌和产品品牌。

四是塑造芒果文化,促进产业提升。要强化乡土技术人才队伍培育和社会化技术服务,探索芒果产业发展与绿水青山、休闲农业等的融合途径,激活乡村振兴内生动力,以产业组织化促进生产标准化,把品牌建设作为产业发展的战略,不断提升芒果产业的质量效益和竞争力,引领产业绿色健康发展。

Introduction

1.1 The overview

The mango (*Mangifera india* L.), a common name for Mangifera, belongs to the Anacardiaceae family and Mangifera. As "the king of tropical fruits", the mango (Figure 1-1) has a global planting area of 6.407 million hm^2, and its output exceeds 56 million t (FAO, 2020),

Figure 1-1　Mango

ranking fifth in the world after citrus, bananas, grapes and apples.

Originated from India, mango is now widely distributed between 30 degrees north latitude and 30 degrees south latitude where tropical and subtropical areas have the coldest average temperature above 11°C winter and absolute low temperature above 3.7 °C. That area is as far north as the south of Chinese Sichuan Province and southern islands of Japan, and as far south as the the south of Africa. Mango is cultivated in more than 100 countries around the world. The main producers are India, Thailand, China, Indonesia, the Philippines, Mexico, Pakistan, Nigeria, Egypt, Cote d 'Ivoire and so on. The main export countries include Mexico, Brazil, Ecuador, Peru, India, Pakistan, Thailand, the Philippines and China, and the main import countries and regions are the United States, Canada, Saudi Arabia, the United Arab Emirates, Kuwait, Japan, Singapore, the United Kingdom, France, Russia, the Netherlands, Belgium, Germany, and so on.

The mango boasts the advantages of wide adaptability, quick growth, strong stress resistance, early fruit (2–3 years after planting), easy cultivation and management, high yield, long economic life (more than 50 years). What's more, it has a beautiful shape, and the juicy pulp enjoys attractive color, smooth texture, rich and unique fragrance and abundant nutrition. That is why mangoes are extremely popular with consumers.

1.2 Nutritional value and function

1.2.1 Nutritional value

Besides delicate meat and sweet smell, the mango has high nutritional value and is rich in protein, fat, carbohydrates, dietary fiber, calcium, iron, phosphorus, potassium, sodium, copper, magnesium, zinc, selenium, manganese, vitamin B_1, vitamin B_2, niacin, vitamin C, vitamin E, vitamin A, carotene, gallic acid, quercetin, mango-ketoic acid, isomango-ketoic acid, albo ketoic acid, albo ketoic acid, mango-ketoic acid, flavonoids, superoxide gum Chemase, β-cryptoxanthin, lycopene, syringic acid, quercetin, etc (Figure 1-2). Incorporating the essence of tropical fruits in it, the mango can help you moisturize the skin, reduce blood pressure and lipid. Thanks to the effect of clearing the intestines and stomach, the mango can prevent vomiting when you get carsickness or seasick.

1.2.2 Health effects

As the mango is cool in nature, sweet and sour in taste, and can be lungs, spleens, stomachs through, it is beneficial to stomachs, thirst, and urinating. The mango is able to not only cure a series of diseases such as thirst and dry throat, loss of appetite, dyspepsia, dizziness and vomiting, sore throat and hoarseness, cough and phlegm, asthma and so on, but help prevent diseases.

Anti-oxidant and anti-cancer: Mango contains a large number of active ingredients such as superoxide dismutase and flavonoids, which have the function of cancer prevention and anti-cancer.

Figure 1-2　Mango pulp

Prevention of cardiovascular and cerebrovascular diseases: with vitamins and minerals, etc, the mango can help prevent arteriosclerosis and high blood pressure. Because the mango contains more vitamin C than ordinary fruits, and its pulp will not be influenced by heating and processing. Regular consumption of mango can supplement the consumption of vitamin C in your body, reduce cholesterol and triglyceride, and is conducive to the prevention and treatment of cardiovascular diseases.

Cough Suppressant & Expectorant: The mangiferin in the mango has the effect of dispelling disease and relieving cough, and has an auxiliary therapeutic effect on cough, phlegm and asthma.

Eye care, beauty and anti-aging: rich in sugar and vitamins, particularly vitamin A which the mango has the most among fruits, it has the effect of brightening the eyes.

Strengthening the stomach and stopping dizziness: The mango has the effect of clearing the stomach and intestines, and has certain anti-emesis effect for carsickness and seasickness.

Antibacterial and anti-inflammatory: The immature mango pulp, bark, stem can inhibit pyococcus, *Escherichia coli*, etc. The extract of mango leaf also has the effect of inhibiting pyococcus, *Escherichia coli*, and can treat human skin, digestive tract infections.

1.3 Edible methods and storage skills

1.3.1 Edible methods

Eating mango directly after ripening is the most common way, which can maximize the retention of the taste flavor and nutritional value. Wash the ripe mango, then run a paring knife lengthwise from end to end, cutting it in half against the pit. Make another cut into the half with the pit to remove the pit and then remove the pulp attached to the pit. Next, before eating, use a knife to make a few more longitudinal and transverse cuts in the mango profile, taking care not to cut the skin. When eating, you can easily eat the delicious mango by pointing it upward with your finger (Figure 1-3).

People in Guangdong, Guangxi and other places like to pickle and eat them. Some people like to peel the raw mangoes and add the chili to flesh.

The mango can be processed into the dehydrated mango, preserved mango, mango paste, canned mango and so on.

Figure 1-3 Golden yellow mango section

1.3.2 Tips for eating mango

Delicious as it is, you are not advised to overeat. The mango is high in sugar content with 12.9 g of sugar per 100g of flesh. It is likely to cause hyperglycemia in case of excessive intake. Therefore, it is not healthy for diabetes patients. In addition, as the mango pulp contains high potassium content, it is not suitable to eat too much after meals or on an empty stomach. Patients with kidney disease are suggested to eat less because mangoes are not favorable to the recovery chronic kidney disease. Mangoes are also inedible for women during their menstrual period.

The mango contains substances stimulating the skin, such as fruit acids, amino acids and various proteins, uronic acids particularly in immature ones. Those can stimulate the skin mucosa and cause allergy. When we eat mango, it is easy to dip the mango juice to the corner of the mouth, cheek and other parts, stimulating the facial skin and causing facial redness and inflammation. In serious cases, eye redness and pain may occur. Some consumers deeply love mango's sweet, smooth taste, while others are allergic to it.

For most people, allergies occurs because of bad habits. Some children or adults used to peel off the skin and bite directly. However, there are many irritants in the mango skin. If you use your hands to peel them off, you should wash your hands in time, or it is necessary to take disposable gloves to handle the peel. You can peel it off a little more when you eat. You can also use the spoon to take to your mouth, so that the mouth and face as far as possible do not contact the mango juice.

If the allergy occurs, you can rinse with ice water to wash away some of the remaining mango juice which is easy to cause allergy. Generally speaking, this kind of allergy will cause redness, swelling and itching, as well as a burning sensation. Rinsing with ice water can calm the skin, reduce the feeling of redness, swelling and burning, and also play a good anti-itching role.

If you have allergic symptoms after eating mango for several consecutive times, it indicates that you are allergic. It is not recommended to continue eating mango. It is also recommended to avoid contact with the trunk, stem and leaves of mangoes, especially to avoid contact with the juice of mangoes' tree.

1.4 Classification and main varieties of mangoes

1.4.1 Classification of mangoes

There are more than 1,000 cultivated varieties of mangoes in the world, including monembryony and polyembryony species. There are more than 40 varieties cultivated commercially in China, which are mainly distributed in Hainan, Yunnan, Guangxi, Guangdong, Fujian, Taiwan and other places.

Monembryony seeds have only one embryo and produce one seedling after sowing. The variability of bearing trees is high and can not maintain good characteristics of the female parent. Indian Mangoes and its bearing offspring (such as red manchus), Chinese Jyhua mango, Guixiang mango, Chuan mango, Yuexi No. 1 mango and Hongxiangya mango are all single embryo varieties. Polyembryonic type mango seeds have multiple embryos, which can produce several seedlings after sowing. The embryos that can develop into seedlings are mostly asexual, and the bearing tree has little variability, and most of them can maintain the maternal traits. The Philippine varieties, the Thai mangoes and the Tu mangoes of Hainan Province belong to this type.

1.4.2 Several Chinese main mangoes varieties

1.4.2.1 Jinhuang mango

It was selected and bred by Mr. Huang Jinhuang from Taiwan Province of China, so it was named Jin-huang mangoes. It was introduced to the mainland China and cultivated in 1999, and now it is mainly distributed in Guangdong, Guangxi, Yunnan, Fujian, Taiwan and other provinces (regions). The weight of the single fruit is as large as 500–1250 g, and the largest can be up to 2,400 g. The mango has long ovate type, thick skin. When the mango grows ripe, it would have small pit, less fiber, more juice and sweet flavor (Figure 1-4).

Figure 1-4 Jinhuang mangoes

Its edible rate is 80.1%, soluble solid 17%–19%, sugar content 17%. What's more, this kind of mango has excellent quality, good commodity nature, pretty storage and transport resistance.

Variety characteristics: strong tree, tall crown, large and sparse flowers, high yield. Generally per mu[①] yield can reach 2,000–3,000 kg, and it has a certain resistance to anthrax.

Shortcomings: prone to have "blister disease" and some physiological diseases.

1.4.2.2 Guifei Mango

Variety source: The Red Guifei, also known as Hongjinlong or Jinfeng, is selected from Taiwan Province of China and has become one of the main varieties in Hainan. It is also planted widely in Yunnan, Guangdong, Sichuan Province.

Variety characteristics: This kind is strong and prosperous. It has round head shape of crown, wide and long leaves, premature and fruitful advantages. The tree begins to bear fruits in 4–5 years, and per plant yield of grafted tree is 20–30 kg or higher. Its fruit is long oval, and has small tip. Fruits' weight varies, and the small ones are about 150 g, the big 500–600 g (Figure 1-5).

Advantages: The fruit is smooth and crimson. When ripe, it has reddish yellow color without any spots and looks gorgeous and attractive. Its pulp is orange, and has small pit, no fiber and

Figure 1-5 Guifei mango

---

① 1 mu≈667 m², 15 mu=1 hm².

sufficient water. Its edible rate reach 74.5%–84.3%, soluble solid of 14%–17%, sugar degree of up to 14–18 degrees, which bring excellent quality. At the harvest period, it is dry and the light is sufficient, the fruit is more tolerant to storage and transportation. The variety is well-known and popular among consumers in the market.

1.4.2.3　Tainong No.1

Selected and bred by Fengshan Horticultural Institute in Taiwan Province, it is now one of the main Manago varieties in Hainan Province.

Variety characteristics: strong crown, strong growth potential, upright, early flowering, long flowering period, strong wind resistance and disease resistance, high ratio of bisexual flowers, stable yield, better resistance to anthrax, wide adaptability. Generally, the yield per plant can reach 5–10 kg or higher 3 years after transplanting.

Advantages: with single weight of 150–200 g, the Tainong No.1 is green in growing period and its sunny slope of shoulder side has the color of carmine. In mature period, the color would turn golden. With beautiful appearance, the flesh is dark yellow, and has fine tissue, sweet taste, less fiber, and smooth texture. The quality is good. The research results of South China Tropical Agricultural University show that Taiwan variety has 16.8% soluble solids, 16.76% sugar, 60.6% edible part, resistance to storage and transportation, long shelf life (Figure 1-6).

Disadvantages: Tainong No.1 has the disadvantages of smaller fruit type and higher labor and production costs than other varieties during bagging.

Figure 1-6　Tainong No.1

1.4.2.4　Kent mangoes

Originated from Florida, USA, it was introduced in 1984 by the Institute of South Tropical Crops, Chinese Academy of Tropical Agricultural Sciences. With ovoid shape, the fruit is about 10.8 cm long, about 10.0 cm wide, about 9.2 cm thick, and its average weight is 447 g. In mature period, its pericarp appears from yellow-green to apricot yellow with dark red at head. With orange-yellow color, its juicy flesh has fine texture, less fiber, rich flavor and sweet taste. It has small seeds and monoembryonic breeding. Fruit ripening period is in July or August. With dense branches, the high-yield variety has advantages of more resistance to storage and transportation and disadvantages of poor wind resistance, which leads to the fact that it is suitable for planting in typhoon-free areas. It is now the main variety in Jinsha River dry-hot valley in Sichuan and Yunnan Province (Figure 1-7).

Figure 1-7　Kent mango

1.4.2.5　Kiett mango

Originated from Florida, the United States, Kiett mango is a late maturing variety. It was introduced by the Institute of South Tropical Crops, Chinese Academy of Tropical Agricultural Sciences in 1984, and is now widely cultivated in Taiwan, Guangxi, Yunnan and Panzhihua, Sichuan.

Variety characteristics: Known for high and stable yield, the Kiett tree has characteristics of medium potential, open branches, late flowering and ripening, and dark green leaves. It has advantages of strong wind resistance and cold resistance. With the oval shape and obvious nose, the fruit has different sizes with average weight of 680 g and more than 2,000 g of big ones. The immature fruit appears gray purple green, and turns dark red with lavender powder when mature.

This variety is especially suitable for growing in typhoon-free areas and high-temperature arid areas in summer, and is one of the main varieties in Panzhihua area of Sichuan Province (Figure 1-8).

Advantages: The fruit has thin skin, small nucleus, thick meat, less fiber, orange pulp with 14%–19% sugar content, about 75% edible rate and slightly sour taste close to the core. Kiett mango is a late maturing variety, and fill the gap in the mango market when Tainong, Guifei, Jinhuang, coconut incense and other early maturing varieties completely exit markets, which brings pretty good market sales.

Disadvantages: In some areas, the yield is not stable enough. Due to the market competition of other late-maturing varieties, the planting area has been gradually declining in recent years.

Figure 1-8　Kiett mango

1.4.2.6　Aiwen mangoes

Tanslated as Aierwen, Ouwen mangoes, the third-generation red mangoes, Aiwen mangoes is originated from Florida, USA. It is also known as apple mangoes when introduced into Taiwan Province of China in 1954. It became the main cultivar in Taiwan. In 1984, it was introduced into South Asian Tropical Crops Research Institute of Chinese Academy of Tropical Agricultural Sciences from Australia. It is now widely planted in Hongge Township, Panzhihua and Changjiang, Hainan.

Variety characteristics: Suitable for dense planting, the tree has relatively short size, round crown, conical inflorescence, light red pedicels, large flower spica, high fruit setting rate. Its flowering period is in January to February, and it generally can produce 3 batches of flowers, which brings stable yield.

Advantages: In immature period, the fruit has obovate shape, purple-green or crab-green color, fuchsia of covering color. The mature fruit has a dark yellow base and a bright red covering, which is known as the ruby among mangoes. The immature pulp is yellow, and has greasy and slippery meat, less fiber, sweet taste and good quality. the mature one tunes golden yellow, and has no fiber, sugar content of 14%–16%, soluble solid content of 15%–24%. Its taste is sweet and excellent, which is popular among consumers (Figure 1-9).

Disadvantages: The yield is greatly affected by the climate.

Figure 1-9   Aiwen mango

1.4.2.7 Haidun mango

Originated from Florida, USA, the average osphere-shaped Haidun mango is about 10.4 cm long, about 9.6 cm wide, about 8.9 cm thick. The single fruit weight is 350 g. the mature peel is orange-yellow with bright red halo. The flesh is orange-yellow, and has slightly coarse texture, less fiber, sweet taste, strong fragrance, and high quality. The fruit's maturity period is from June to July, and the seed is single embryo. This variety is suitable for cultivation in high temperature, drought and sunny places, and is unstable in yield and color if cultivated in rainy areas. The fruit has strong resistance to storage and transportation. This variety is also the main variety in South Africa, Mexico, Israel and other countries (Figure 1-10).

Figure 1-10　Haidun mangoes

1.4.2.8　White ivory mango

Originated from Thailand, the white ivory mango is one of the country's main export varieties and is now one of the main commercial cultivars in Hainan and Yunnan Province.

Variety characteristics: The white ivory mango has a strong tree potential. This kind of tree has high laurel and round head shape, and its branch is upright with small branches. Flowering is from late March to late April. Inflorescences is conical, and its rachis is reddish. The fruit is large, long oval, curved, and has fruit mouth marks. Like nascent ivory, the fruit is therefore named as the white ivory mango (Figure 1-11).

Advantages: The average weight of single fruit is 680 g, and the larger one can reach more than 2,000 g. The mature fruit is golden and has thin skin, small core, thick meat, and less fiber. Its orange flesh has the sugar of 17% and the fresh and juicy flesh is as sweet as honey. The fruit is more resistant to storage and transport.

1.4.2.9 Nanduomei mango

Originated from Thailand, the Nanduomei mango is also called as Qingpi mango, Baihua mango. It is widely cultivated in Yunnan and Hainan Province.

Variety characteristics: This variety has medium to strong tree potential, leathery and alternate leaves; It has small, yellow or reddish flowers in terminal panicles.

Advantages: With kidney shape, the fruit has distinct groin, and its ripe pericarp is dark green to yellowish green. With thin skin, the delicate and juicy flesh is light yellow to milky yellow, and has the fragrance of honey, little fiber. Its seed is flat and thin and the embryo is polyembryonic. The single fruit weight is 200–300 g with edible part accounting for about 72%. The excellent quality makes this fruit as the ideal fresh food variety (Figure 1-12).

Disadvantages: In low-temperature rainy period, the variety will appear flowers and not fruiting, and is prone to dehiscence, leading to medium yield. Its plants is susceptible to gummosis. Due to the cyan peel and pale flesh color, its sales and value would be affected in some places.

1.4.2.10 Jyhua mango

The Jyhua mango, selected from the offspring of Thailand mangoes by the College of Agriculture of Guangxi University, is now planted widely in Guangxi and Guangdong

Figure 1-11　White ivory mangoes

Figure 1-12　Nanduomei mangoes

Province.

Variety characteristics: The variety has medium tree potential, open branches, early bearing, high yield, pretty stable yield. The grafting seedling can bear fruits in 3–4 years. A six-year-old tree has the yield of 1,000 kg per mu or higher. The Inflorescence is conical and rachis is purplish red. The variety has late flowering of from late March to April, which can avoid low temperature rainy periods. The Jyhua mango is more resistant to pruning, suitable for dwarf and dense planting.

Advantages: The fruit is obliquely long oval, sharp at both ends. Its rind is grey-green, and the sunny side is reddish yellow. It turns to bright yellow and has thick wax powder after ripening. The weight of single fruit is 250–300 g. The juicy fruit has beautiful shape, yellow flesh, smooth flesh, moderately sweet and sour taste, light fragrance, little or no fiber. It has edible rate of 64%–73%, soluble solids of 13%–15% and 12–15 g/100 mL of sugar, acid content of 0.09–0.65 g / 100 mL. It is resistant to storage and transportation (Figure 1-13).

Figure 1-13　Jyhua mango

Disadvantages: This variety is sensitive to low temperature and is not cold resistant, so it is susceptible to cold injury.

1.4.2.11　R2E2

R2E2, also known as Australian mangoes, is originated from Australia. The new variety was introduced by the Institute of South Tropical Crops of the Chinese Academy of Tropical Agricultural Sciences in 1997. The average fruit weight is 716 g and the quality is excellent.

Variety characteristics: This variety has strong viability, early bearing and high yield and is easy to manage.

Advantages: Australian single fruit weight is 500–1,500 g. Similar to the shape of apple, the fruit has smooth and beautiful appearance, golden color with red halo, rich fragrance, delicious meat, sweet taste and no fiber (Figure 1-14).

Figure 1-14   R2E2

1.5   Mango trees' morphological characteristics and growing environment

1.5.1   Morphological characteristics of mango trees

Mango trees belong to large evergreen type and is 10–20 m tall. Their glabrous barks show grayish brown, and branchlets are brown. Variable in shape and size, their leaves are usually oblong or oblong-lanceolate and is 12–30 cm long and 3.5–6.5 cm wide. The thinly leathery leaves often cluster at the branch apex. With various shape of flat to acuminate, long acuminate or acute acuminate, their glabrous leaves have cuneate or suborbicular base, undulate margin, light shine. Raised on both surfaces, their lateral veins have 20–25 pairs and is obliquely ascending. Their reticulate veins are not visible, and petioles are 2–6 cm long with upper grooves and expanded bases.

Mangoes' densely distributed flowers have panicles and is 20–35 cm long and grayish-yellow puberulent. Their branches are open and 6–15 cm long at the base. The samll bracts are lanceolate, about 1.5 mm long, puberulent, hybrid, yellow or pale yellow. The pedicels are 1.5–3 mm long. The glabrous petals are oblong or oblong-lanceolate, 3.5–4 mm long, about 1.5 mm wide with 3–5 brown raised veins.They are outer rolled when flowering and disc chucks dilate and are fleshy. Only one stamen develops and is about 2.5 mm long. Anthers are ovate and glabrous ovaries are oblique ovate and about 1.5 mm in diameter. The styles are subterminal, and about 2.5 mm long.

The shape of mangoes is oval, kidney or obovate shape. Mature fruits' skin is green, yellow or purplish red, and the flesh is yellow or orange. The juice and fiber vary from variety to variety.

Drupes are large, kidney-shaped (the shape of different cultivars varies greatly), flattened, 5–10 cm long, 3–4.5 cm wide, yellow when the fruit is ripe. Their flesh is hypertrophic, bright yellow, sweet and has hard pit.

1.5.2 The growing habit of mango tree

Shoots' growth habits. The shoots of the mango branch are of the afro-subtype growth, and the buds are wrapped by bracts. In growing periods, the bracts first break open, and the bud tips extend. As the leaves develop, the bracts fall off immediately. The middle and lower leaves are alternate, and the leaf spacing is large. Generally, seedlings and young trees take 6 to 8 times sprouting a year, young bearing trees 2 to 4 times, and adult trees 1 to 2 times. The shoots sprouting in March to May are spring shoots. Summer shoots sprout in June to August, autumn shoots from September to November, and winter shoots from December to February. In Hainan,

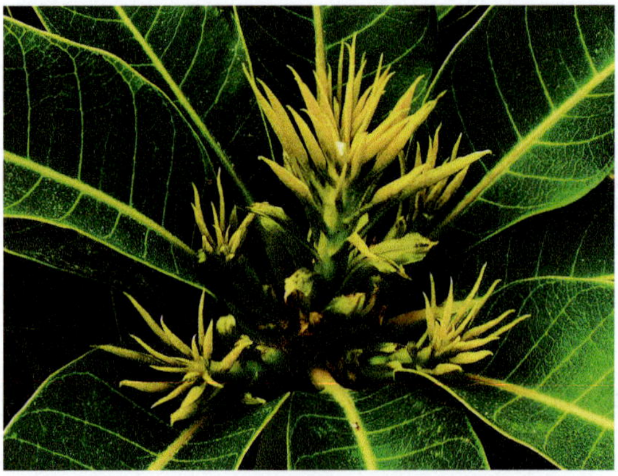

leafbud

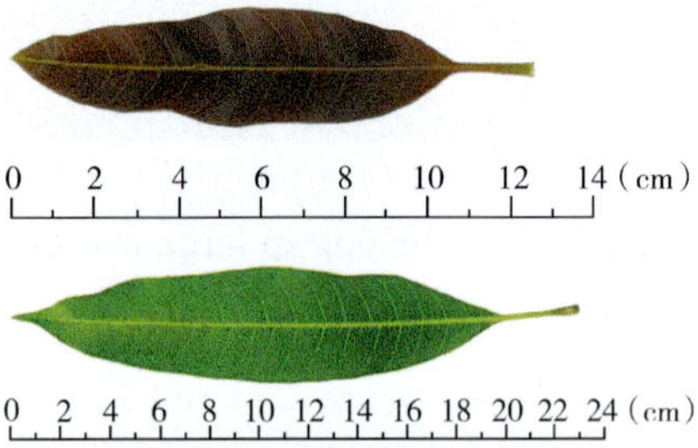

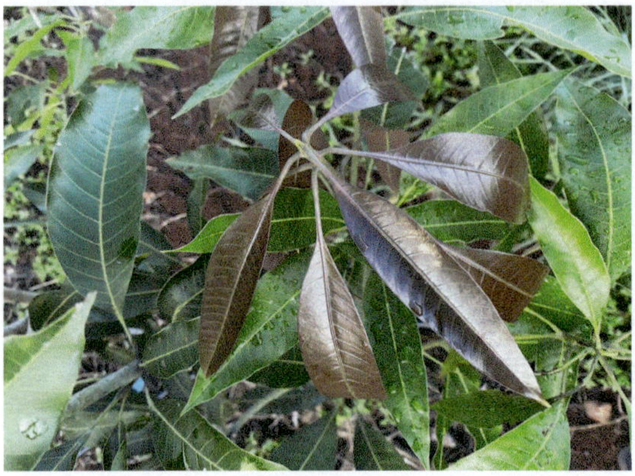

tender leaf

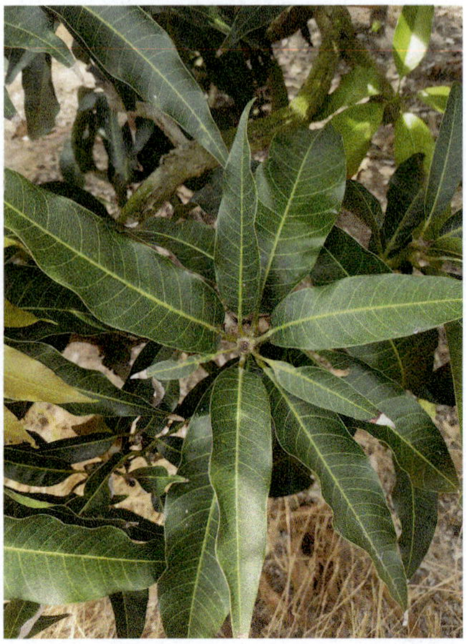

mature leaves

Figure 1-15 Leafbud, tender leaf, mature leaves

the autumn shoot is the main bearing parent branch, and the spring and summer shoot can also be the bearing parent branch. Under good conditions, winter shoots of some varieties can blossom and bear from December to January of the next year. It lasted 15–35 d from bud germinating to shoot stopping growth and leaf aging. The duration of summer and autumn shoot is shorter, while the duration of winter shoot is longer. Shoot growth alternates with root growth (Figure 1-15).

Flower buds differentiate. Under normal conditions, mango flower bud differentiation starts from late October to November. The flower bud differentiation is not restricted by the season by the use of medicine to promote the flower. It took 20–39 d for the first flower to open from the bud to the inflorescence, but the inflorescence could continue to extend after the first flower opened. Moderate low temperature drought is conducive to flower bud differentiation, high temperature is conducive to bisexual flower formation.

Bloom. Mango trees naturally bloom from December to February, sometimes as early as November or as late as March, and bloom around the Spring Festival. It takes 15–25 d for a single inflorescence to open from the first flower to the whole inflorescence, and about 50 d for a single tree (Figure 1-16). Mango flower is divided into bisexual flowers and male flowers, bisexual flowers have normal development of stamen and pistil, can carry out normal pollination

Figure 1-16　Mango flowers

fertilization and fruiting, male flowers without pistil, after flowering can not bear. In most cultivars, bisexual flowers account for more than 15%. A flower spreads from petal to stigma dry for about 1.5 d.

The fruit. The ovary began to expand after flowering and fertilization, and rapidly increased about 1.5 months later, but the growth was very slow or no growth 10–15 d before fruit picking. At this time, the main reason was fruit thickening, fullness and weight increase. From flowering to ripening, it takes 85–110 d for early ripening varieties, 100–120 d for medium ripening varieties, and 120–150 d for late ripening varieties (Figure 1-17). During fruit development, there are two obvious peak fruit falling: the first is about two weeks after anthesis, mainly the small fruit with poor fertilization is yellow and falls off, and the second is 4–7 weeks after anthesis, except for a small part of malformed fruit or aborted fruit, most of the fruit falling is caused by lack of nutrients and water. Physiologic fruiting rarely occurs 2.5 months after flowering. The fruit harvest period of conventional cultivars in China is from May to July, which varies according to varieties and regions.

Figure 1-17　Mango sets after flowering

1.5.3　Environmental requirements for growth

　　Temperature. Mangoes like warmth in nature and dislike frost. The optimum growth temperature is 25–30 ℃. below 20 ℃ the growth would be slowed. Below 10 ℃ the leaf and inflorescence will stop growing, and the nearly mature fruit will be affected by cold. The average annual temperature of mango production area is above 20 ℃, and the lowest monthly average temperature is above 15 ℃. Low temperature would lead to poor pollination and fertilization, even embryo abortive death. When the temperature is higher than 37 ℃, the florets and fruits will produce sunburn. When the temperature is lower than 10 ℃, the new shoots and flower spikes will stop growing. When the temperature is lower than 5 ℃, the seedlings, tender shoots and flower spikes catch a chill. When the temperature is about 0 ℃, the flower spikes, tender shoots and peripheral leaves of adult trees will be affected and die in severe cases. Below -3 ℃, young trees freeze to death and big trees are severely frozen.

　　Light. Mango trees love light and sufficient light can promote flower bud differentiation, increase flowering and fruit setting rate and fruit quality. Usually, in empty environment or the sunny side of crown, mango trees would be in full bloom, have more branches and leaves, closed canopy, and high fruit setting rate. In insufficient light, mango trees have fewer blossoms, fruits, and poor fruit appearance and quality. The light transmittance condition in the garden and tree can be improved by shaping and pruning to increase yield and prolong the perinatal period (Figure 1-18).

Figure 1-18　Mangoes orchard

Water content. Mango trees grow well in areas with annual precipitation of 700–2,000 mm. If the air is too dry during flowering period and early fruiting period, it is easy to cause flower and fruit abscission. If the rain is too much, it is easy to cause rotten flowers and poor pollination and fertilization. if the summer rainfall is too concentrated, it often causes serious fruit diseases. The autumn drought after harvest will affect the germination and growth of the mother branches in autumn shoots.

Soil. Mango trees does not have high requirements on growing soil. The deep soil layer, the less than 3 m water table, good drainage, and the slightly acidic loam or sandy loam is favorable to growth.

1.6 Selection and storage methods

1.6.1 How to choose mangoes

First, the hardness. If you can feel the pulp sag when pressing the mango's skin, proving it is mature. If not, the mango is not ripe.

Second, the color. The skin color of most varieties of mangoes is orange yellow when they get ripe, and green when not ripe. When we cut the mango, if the core is hard and the flesh is yellow, proving it is ripe. If the core is soft and the flesh is white or green, it is not ripe.

Third, the specific gravity. We can put the mango into water to test. If it sinks, it is ripe. If it floats to the surface, it is not ripe (Figure 1-19).

Fourth, the smell. If the mango has sweet and sour taste, it is ripe. While it is green or tasteless, it is not ripe.

Figure 1-19　Guifei mangoes

1.6.2　How to distinguish mangoes between natural ripening and forced ripening

First, the color. The color distribution of natural ripening mango is uneven. Forced ripening mangoes have the small green head tip, while other parts of the skin are yellow.

Second, the fragrance. Naturally ripened mangoes can mostly smell a fruity fragrance, and forced ripening mangoes smell light or bad.

Third, the hardness. Naturally ripened mangoes have better elasticity, while forced ripening mangoes are softer overall.

Fourth, the eating experience. When you eat the forced ripening mangoes, there is not much water and sugar. If you tear its skin directly, you will feel the skin is tight to the flesh. It is quite difficult to tear into pieces. Natural ripening mangoes have rich juice and high sweetness. It is easy to separate peel from the flesh. Mangoes with high maturity have less allergenic components, so it is not prone to allergic phenomena and other physical discomfort symptoms after eating.

Fifth, the powder layer. Natural ripening mangoes have a layer of gray powder on the surface of the skin, and the forced ripening mangoes do not have this powder layer.

1.6.3　How to store the mango

As a respiratory climacteric fruit, the mango is suitable for harvesting medium well. At this time, the mango is hard and astringent, and can only be eaten after post-ripening (that is, the starch contained in the pulp is converted into soluble sugars). The storage period at normal temperature is generally only 7–12 d, and the storage at moderate low temperature can extend the freshness period. The cold storage at 10–13 ℃ is commonly used on the market. When the temperature is lower than 10 ℃, the mango is prone to suffering from cold injury and cannot be post-ripened. So if not ripe, the mango should not be refrigerated at too low temperature. Otherwise it will not ripen and soften. When soft and ripe, it can be stored in the refrigerator. But it doesn't last long, because even refrigeration can't stop post-harvest diseases. Anthracnose and root rot are two major diseases after harvest. In the middle and late period of storage, the black plaques on the peel are the manifestations of post-harvest diseases. The plaques will expand continuously with the extension of time, and even the pulp will rot and exude black and brown juice.

It is better for the mango to be preserved in a cool, ventilated and dry environment away from light. When picking fruit, you are advised to leave 2–3 cm fruit stalks. The branches of mango should be kept dry and complete as far as possible to avoid pectin sticking to the fruit.

If you want the mango to ripen and soften quickly, you can place it with a ripe banana and the ethylene gas released by the banana to ripen the mango faster.

The mango can be made into jam after ripening, or you can peel off the skin and frozen it in the refrigerator. That method can preserve the mango for a long time. You can enjoy a distinct mango taste after thawing.

Dry and store. For uneaten ripe mangoes, you can use freeze-drying or sun-drying to remove excess water from the mangoes, and then seal in dry environment. The dried mangoes can be stored for more than 6 months.

Global Mango Industry Pattern and Chinese Mango Production

2.1 Global mango industry pattern

Mango is one of the world's five largest planted fruits, with its output second only to grapes, citrus, bananas, apples, and production scale ranking the third. According to the Food and Agriculture Organization of the United Nations (FAO), there are 103 countries producing mangoes, mainly in Asia, South America and Africa, covering from south Sichuan province to South America, across the region between 30° north and south latitude. Asia is the region with the largest mango planting area, and the total output of mango accounts for about 85% of the world's mango output. It is followed by the Americas, which produce about 14% of the world's total. The ripening seasons of mangoes in major mango producing countries in the world are shown in the table 2-1. The top 10 countries for mango production in 2020 are as follows.

India: India is the country with the largest mango harvest area in the world, with a harvest area of nearly 2 million hm^2. It produces about 16,337,400 t of mangoes every year, accounting for 42.2% of the world's total mango output. India has a wide variety of mangoes and is considered to be the country with the largest mango species.

China: China is the world's second largest mango producer. By 2020, the mango planting area in China has reached 349,400 hm^2, with a total output of 3.306 million t, accounting for 8.75% of the global total output, with an output value of 20.52 billion yuan.

Thailand: The annual production of mango is about 2,550,500 t, accounting for 6.5% of the world's total mango production, ranking the third in the global mango production.

Pakistan: Annual mango production is about 1,784,300 t, accounting for 4.6% of the world's total mango production, ranking the fourth in the global mango production.

Mexico: The annual mango production of about 1,632,700 t, accounting for 4.2% of the world's total mango production, ranking the fifth in global mango production.

Indonesia: Annual mango output is about 1,313,500 t, accounting for 4.1% of the world's total mango production. Indonesia's main mango producing areas include East Java, South Sulawesi Island, East Kalimantan, West Nusa Tenggara Province. Indonesia is the sixth largest mango producer in the world.

Brazil: The world's seventh largest mango producer, the mango annual output is 1,188,900 t, accounting for 4% of the world's total mango production.

Bangladesh: The eighth largest mango producer in the world, producing about 1,047,900 t of mangoes per year, accounting for 3.9% of the world's total mango production. Due to its favorable climate, the Bangladesh market provides fresh and juicy mangoes almost all year round.

Philippines: The ninth largest mango producer in the world, producing about 823,600 t of mangoes per year, accounting for 3.6 percent of the world's total mango production.

Nigeria: The tenth largest mango producer in the world, producing about 790,200 t of mangoes per year, accounting for 3% of the world's total mango output.

Table 2-1 Mango ripening season in the world's major mango producing countries

Area of Production		Market Time
China	Hainan	March to May (early fruit in southern Hainan from November to March of the next year, some late fruit in June)
	Guangdong	Mid-June to Mid-August
	Guangxi	May to September (concentrated picking in June and July)
	Yunnan	May to August (small amount of picking from September to November)
	Panzhihua	August to November (early fruit has been produced from June to July)
Israil		August–September
Brazil		September to December
Peru		January to February
Australia		December to February of the following year

续表

Area of Production	Market Time
South Africa	January to March
Philippines, Thailand	January to May
Vietnam	September to June of the following year (peak in February – March)
Mali, Cote d 'Ivoire, Burkina Faso	March to May
India, Mexico, Venezuela, Puerto Rico, Costa Rica	May to July (peak period)

2.2 Production situation of some main mango producing countries

2.2.1 India

Mango has a cultivation history of more than 4,000 years in India which is recognized as the earliest recorded mango cultivation country with the world's largest mango cultivation area as well as the highest output. In 2016, the cultivated area of Indian mango was about 2.12 million hm^2, the yield was 19 million t, and the yield per unit area was 8.9 t/ hm^2. Mangoes are produced in 14 Indian states, in order of cultivation area from largest to smallest: Andhra Pradesh, Uttar Pradesh, Orissa, Karnataka, Telangana, Tamil Nadu, Maharashtra, Gujarat, Bihar, West Bengal, Chhattisgarh, Kerala, Jharkhand, Madhya Pradesh. The highest yield per unit area was in Uttar Pradesh, reaching 17.14 t/ hm^2. The lowest yield per unit was in Maharashtra, which was only 3.28 t/ hm^2. Hundreds of mango varieties are grown in India, and the harvest season is from March to August. The main commercial varieties include Alphanso, Amrapali, Aswina, Banganpalli, BombayGreen, Chausa, Chenlkurasam, Chinnarasam, Dashehari, Fazli, Gulabkhas, Himayuddin, Himsagar, Kalpadi, Caesar (Kesar), Kishen Bhog, LakshmanBhog, Langra, Mallika, Mulgoa, Mundappa, Neelum, Pairi, Poiri, Rajapuri, Ramkela, RaniPasanddegn Rataul, Rumani, Sepia, Sukul, Sunderij, suvarnarekha, Totapuri, vanraj and zardalu, among which Alfonso mango planting area is the largest and with the highest yield.

In terms of breeding, India has bred 28 varieties like Mahmood Bahar and 7 varieties such as Pusa surya from hybrid progeny materials. Indian mango seedlings are mainly propagated by grafting, which can be divided into branch grafting and grafting. High-density planting techniques including dwarfing rootstock, canopy control, fertilization and the use of polybulobuzole were used to achieve early high yields. For example, the planting density of Amrapali was 1,600 plants /hm^2, and that of Dashehari was 3 m×2.5 m.

On the tree body management side, India mainly controls tree height and tree potential through pruning techniques and the use of polylobulozole. After pruning, some varieties still maintain high and stable yield, and some have obvious phenomenon of large and small year. Banganpalli, Suvarnarekha, Neelum and Banglora varieties are sensitive to pruning and annual or alternate pruning can control the canopy and open the canopy to benefit fruit setting and improve fruit quality. Polylobuzole is currently widely used in India to control tree growth and promote flower formation, usually by soil leaching and foliar spraying, soil application concentration is 2.5–10 g per tree, leaf spraying concentration is 0.5–2.0 g/L, mainly depending on the age and tree growth. The soil leaching effect is better, and the application time is from July to October. In fruit preservation, naphthoacetic acid and gibberellin are mainly adopted. The recommended concentration was 20 mg/kg for Mallika and 10 mg/kg for Dashehari. The early planting of peanuts, ginger, turmeric, pineapple and potatoes can be used for the intercropping. In the aspect of preservation, low temperature storage (temperature of 7–12 ℃, relative humidity of 85%–95%) and gas storage are mainly used at present. In fruit fly control, fruit fly traps are mainly used to trap fruit flies.

2.2.2 Thailand

Located on the Indo-China Peninsula, Thailand enjoys the reputation of "fruit Kingdom". Thailand is a major producer and exporter of mangoes across the world. The climate conditions are very suitable for mango production and there are many varieties of mangoes. The production and cultivation technology of Thai fruit farmers is high, and the regulation of mango perinatal period has become a widely used technology of the mango production in Thailand. Due to the high yield of mango planting, mango among all economic fruit tree species has the largest cultivation area and is favored by a large number of fruit farmers . In 2016, the cultivated area was 340,000 hm^2 and the yield was 3.13 million t, of which 96% was used for fresh food, 1.17% for processing and only 2.23% for export. In 2017, Thailand exported 56,000 t of mangoes, mainly to South Korea, Japan, China and European countries in Asia.

Mangoes are commercially produced throughout Thailand. The main variety Namdokmai is mainly cultivated in northern Thailand (Phuthiluo Province, Chiang Mai Province), northeastern

Thailand (Nakhon Ratchasima Province, Lai Province), eastern Thailand (Chachunsao Province, Saxiang Province) and southern Thailand (Bashu Province).

Naturally, Thailand's primary mango harvest is from April to May. The growing season of Thai mangoes coincides with the dry season in Thailand, so the fruit quality is very good. The ripening time of Thai mangoes varies according to latitude, starting from the central region, followed by eastern, northern, northeastern and northern alpine regions. In addition, out-of-season production techniques developed in Thailand in recent years enable annual mangoes to be harvested.

Thailand has more than 200 mango varieties and about 10 commercially cultivated varieties. According to eating habits, mangoes in Thailand can be divided into three main groups. The first category is raw mangoes, which can be harvested and eaten at an unripe stage and have a distinctive crisp texture with no acidity or a slightly sour taste. Raw mango varieties are Khieu Savoi, Raet, Falon and so on. The second category is cooked mangoes, which are harvested when fully ripe and eaten after ripening and softening. The varieties include Namdokmai, Namdokmai Si Thong, Mahachanok, Okrong and so on. The third type is processed mango, which is used for processing into canned mango, dried mango, mango roll, pickled mango, mango juice, mango ice cream, etc. The popular varieties are Kaeo, Mahachanok, Chokanan and so on. In recent years, Thailand has gradually introduced mango varieties with bright skin color such as R2E2 and Yuwen from Australia, Taiwan and other regions for trial planting and cultivation, among which R2E2 has gradually expanded its cultivation area due to its good fruit appearance, fruit flavor and yield.

Thailand usually adopts planting grafting for mango production, while commercial seedling production mainly adopts grafting. By grafting, the scion and rootstock retain their roots at the same time, so the graft survival rate is higher. Another way is to plant 1 to 1.5 years of seedlings in the orchard before grafting, which is cheaper than buying grafted seedlings.

The row spacing of Thai mango plants in conventional cultivation is generally 10 m×6 m, and the planting pit size is generally 30 cm×30 cm×30 cm. In the middle of the area with high groundwater level, fruit farmers usually adopt the way of deep furrow and high furrow planting, the general furrow width is 6–8 m, the furrow width is 1–1.5 m. Generally, 2.5 m × 2.5 m is used in dense cultivation. This dwarf dense planting method is beneficial to pruning, drugging, fruit-thinning, bagging, harvesting and other operations, and the yield per unit area is higher than the conventional cultivation mode.

Pruning is usually done after fruit picking. Pruning back the bearing branches is about

60 cm, and the old and weak branches are cut off to make the tree ventilated and transparent. Farmers often place pruned branches at the roots as mulch. During the flowering period of mango trees, when the inflorescences are 2–3 cm long, farmers apply insecticides to control pests and stop applying insecticides during the blooming period to facilitate insect pollination. Insecticide application was resumed at the beginning of mango setting and the bagging stage. After 30–45 d of mango fruit setting, fruit thinning should be carried out when the fruit size is about 5 cm long and about 2 cm wide, and the fruit that is too small, deformed or damaged by pests and diseases or aborted should be removed. One inflorescence usually retains 1 to 3 fruits to ensure that the fruit size is moderate. When the fruit is about the size of an egg, the farmer bagges it. Apart from the effect of fruit fly control, bagging can also increase the color of the skin, the most commonly used bag is a yellow brown paper bag with a black coating inside. To save costs, fruit farmers usually recycle paper bags for reuse.

In Thailand, mangoes are produced out of season in two ways. One is to use the variety flowering many times a year for production, and the other is to urge flowering in chemical way. At present, Paclobutrazol (Paclobutrazol) has been widely used in Thailand to promote mango flower out of season. Prebulobuzole is usually applied in soil, removing weeds from tree pans before application and keeping the soil moist to improve absorption efficiency. Growers will adjust the amount of prebulobuzol used according to the health of the tree, usually 10 g/m crown diameter. Trees that use too much buzole need to be reduced the dose the following year because the plants can absorb the previous year's residue from the soil. The amount of agent residues in the soil is related to soil texture and is usually lower in well-drained loam than in clay.

Thai growers often apply organic and chemical fertilizers to mango trees. Organic fertilizer is usually applied once a year after harvest and the amount of fertilizer is determined according to the size of the tree. Fertilizer is usually applied twice a year, once with organic fertilizer after the fruit is picked and once as topdressing at the young fruit stage. The amount of fertilizer applied is determined by the age of the tree and the amount of fruit. In addition, growers use foliar fertilizers such as seaweed extract, calcium and boron. Mango trees need a lot of water to grow, especially during the flowering and fruit development stage, which is usually the dry season in Thailand, so farmers need to irrigate their trees. Some growers keep the soil moist by mulching trees.

The main mango pests and diseases in Thailand are Scirtothrips dorsalis Hood, Idioscopus clypealis Lethierry, and Hy pomeces squamosus, Deporaus marginatus, Bactrocera dorsalis Hendel, etc. The main diseases are Colletotrichum gloeosporioides Penz, powdery mildew (Oidium mangiferae Berth), soot disease (Meliola mangiferae Earle) and so on. Some fruit farmers in Thailand use pest monitoring technology to control the occurrence of diseases and pests, such as using sticky plates to monitor the pest population density, and then determine the timing and dosage of control, and also use other physical control methods such as solid bagging. To reduce pesticide use, fruit growers also use sex attractants to control fruit flies and to reduce fruit fly populations.

April to June is the natural mango harvest season in Thailand. With the widespread use of out-of-season flower urging technology, mangoes can be produced all year round in Thailand. Namdokmai fruits are usually harvested 105–115 d after flowering or 55–60 d after bagging. Some growers also judge the timing of mango harvest by its size, shape and skin color.

Mango orchards where the fruit is exported, usually harvested when the fruit is 85% ripe. During harvesting, workers pick the fruit carefully to avoid bruising. Keep long fruit handles and bags temporarily, and transfer them to the packing shed immediately after picking. In the packing shed, first cut the fruit stem to 3–5 cm long, to avoid the fruit pulp at the base of the stem flow out and contaminate the fruit surface, and then remove the fruit bag. The fruit will then be graded to select those that meet export quality requirements and the rest will be re-graded for domestic sale.

Mangoes that meet export quality requirements undergo additional processing to meet export requirements. Wash the skin, remove stains and dust, and visually inspect to make sure there is no anthrax, pests, blemishes or abrasions before removing the stalks completely. This is followed by steam heat treatment, then blow drying in a cool place, and finally packaging. A final quality check will be made before the fruit is crated and transferred to a refrigerated truck.

2.2.3 Pakistan

In 2016, Pakistan's mango planting area was 167,700 hm^2, ranking 7 th in the world. Its output is 1.61 million t, ranking the 6th in the world, with the unit yield of 9.57 t/hm^2. Mango is the second largest fruit crop in Pakistan, mainly exported to the Middle East, Britain, Afghanistan, Southeast Asia and so on. The mango harvest time in Pakistan is from May to October, and the concentration period is from June to July. It is mainly grown in Punjab and Sindh provinces, which account for over 90% of Pakistan's mango production. There are over 100 varieties of mango grown in Pakistan. The main commercial varieties are Anwar Ratole, Baganapalli, Dashehari, Fairi, Gulab Khas, Kala Chaunsa, Langm, Malda, Sindhri, Siroli, Summar Bahist Chaunsa, Suvarnarekha, White Chaunsa, et al. Among them, Summar Bahist Chaunsa has the largest planted area.

Pakistan mainly uses local single embryo varieties as rootstocks, and the cultivation and management techniques are similar to those in India. Mango pests in Pakistan mainly include mango cicada, noctumoth, fruit fly, mealybug, weevil, scale insect and mango gall midge, etc. The most important pest is mango fruit fly. The main mango diseases in Pakistan are black spot, anthrax and powdery mildew. Pakistan has suffered serious loss of mangoes after harvest, which has reduced the quality and price of mangoes. Post-harvest losses are mainly caused by unreasonable post-harvest management, malignant early mining driven by contractors' interests, and inadequate transportation and storage facilities. Because mangoes tend to decay, if mango vendors do not timely sell or process into products, it will result in the backlog of fresh mangoes damaged. Seventy-five percent of mango vendors in Pakistan use traditional packaging methods and local chemicals to ripen their mangoes, causing damage. For the mango export market, Pakistan has adopted the following measures after mango harvest: ① Improve the shelf life of mangoes by post-harvest treatment measures such as fruit washing/waxing, hot water soaking and cold treatment; ② Establish mango export processing zone, export by container; ③ Conduct pre-export inspection; ④ Send sales staff to new markets.

2.2.4 Vietnam

Mangoes are one of the main tropical fruits grown in Vietnam, second only to bananas. Vietnam is the 13th largest mango grower in the world. In 2020, the total area of mango planting in Vietnam reached 87,000 hm^2 and the total output reached 893,200 t, a year-on-year increase of 6.5%. Of these, the Kowloon River Delta accounts for 48%.

In 2020, Vietnam exported more than $180 million worth of mangoes, accounting for 1.15% of the world's total, according to the Agriculture, Processing and Marketing Development Bureau of Vietnam's Ministry of Agriculture and Rural Development. The main export market of Vietnamese mangoes is China (accounting for 83.9%, with export value of $152 million), followed by Russia, the United States, the Republic of Korea, the European Union, Australia and Japan. The Ministry of Agriculture and Rural Development has set a target of 140,000 hm^2 of mango planting area, 1.5 million t of mango production, 650 million US dollars of export volume, and 70% of fresh processing plants reaching advanced technology by 2030.

2.3 Mango industry in China

2.3.1 Distribution of mango industry in China

Mango prefers warmth to frost. The optimum growth temperature is 25~30 ℃, the growth is slow below 20 ℃, when below 10 ℃ the leaf and inflorescence will stop growing, and the nearly mature fruit will be affected by cold.

China is the world's second largest producer of mangoes, with rich variety resources and extensive planting areas. Chinese mangoes are mainly produced in Hainan, Guangdong, Guangxi, Yunnan, Sichuan, Fujian, Guizhou and other places. Among them, Hainan Province is the earliest and generally mature from March to June every year, followed by Guangdong Province from May to August, Guangxi from June to September, Yunnan from May to November, Sichuan and Guizhou from July to October, and Fujian from August to October.

By 2020, China's mango planting area has reached 349,400 hm^2 (5.241 million mu), with a total output of 3.306 million t and an output value of 20.52 billion yuan. In the past 10 years, the yield of mango has maintained a rapid development. In 2011, the yield of mango is only 100.34 million t, and in 2020, the yield of mango is 3.306 million t, which is more than three times that of 2011, with an increase of 2,302,600 t, increased by 229.48%, and with an average annual compound growth rate of about 14.17%.

Currently, mangoes are cultivated in tropical and subtropical areas, and the main production areas are distributed in more than 100 cities and counties in Hainan, Guangdong, Guangxi, Yunnan, Guizhou, Fujian, Sichuan and Taiwan provinces. Tainong No. 1, Jinhuang, Kight, Guremang No. 82, Guifei, Guremang No. 10 and other varieties have wide adaptability and excellent quality, and have become the main varieties of mango industry in China. However, at present, the self-bred varieties only accounted for about 15% of the mango planting area in China, which resulted in serious homogeneity and little differentiation. More imported varieties and less cultivated varieties have become the deep-seated reasons restricting the development of mango industry.

2.3.2 Characteristics of mango industry in China

The domestic mango industry can be divided into five advantageous industrial belts, namely, the early-maturing mango industrial belt in Hainan, the early and middle ripe mango industrial belt in Leizhou Peninsula of Guangdong, the middle ripe mango industrial belt in Youjiang Valley of Guangxi, the mango industrial belt in Southwest Yunnan-South Yunnan-Zhongyuan River Basin and the late ripe mango industrial belt in Jinsha River Dry and hot Valley basin. The dominant varieties developed in different producing areas are also different.

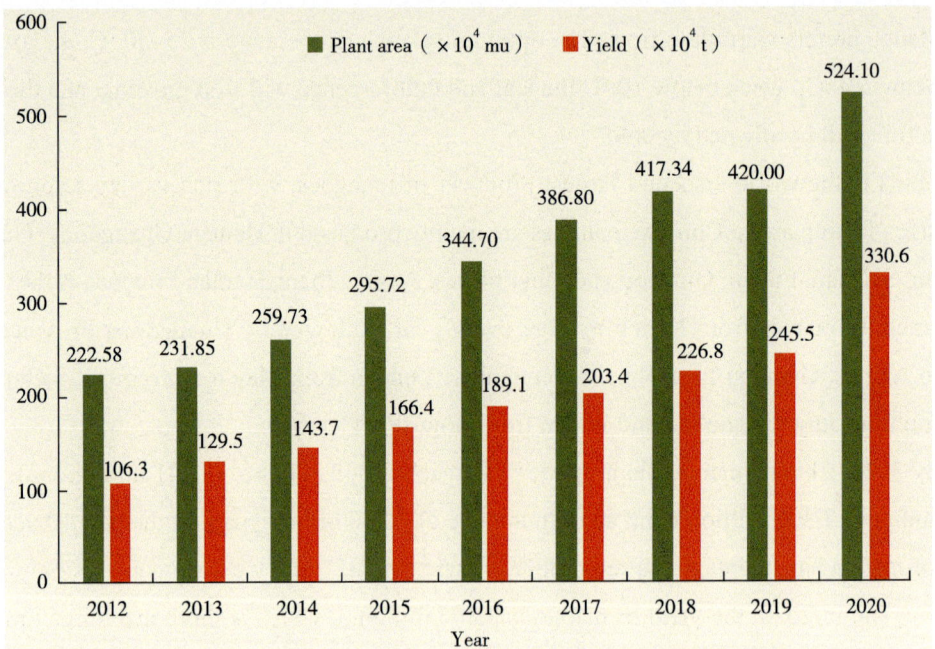

Figure 2-1 Mango planting area and yield in China from 2012 to 2020

2.3.2.1 Hainan early ripe mango predominance industrial belt

The planting area is about 874,800 mu.

Variety structure: Suitable for all mango varieties. The mainstream varieties are Tainong, Jinhuang, Guifei, carbuncle, R2E2 and so on.

Listing time: November to the end of May of the following year, concentrated listing period in March to May each year.

Distribution overview: The largest planting area is Sanya City, which will reach 365,400 mu in 2020. The second is 216,700 mu in Dongfang City; Ledong County ranked third with 170,400 mu, Lingshui county with 60,000 mu and Changjiang county with 37,800 mu.

Regional brand: "Sanya Mango", National Geographic Indication product of China, was selected into the list of advantage areas of agricultural products with Chinese characteristics in 2018; In 2019, it was selected into the China Agricultural Brand directory; In 2020, it was selected into the second group of protected geographical indications between China and Europe.

2.3.2.2 Guangdong Leizhou Peninsula early and medium ripe mango predominance industrial belt

The planting area is about 300,000 mu.

Variety structure: Tainong No. 1, Jinhuang, Guifei, Hongmao, etc.

Available: Mid-June to early August.

Distribution overview: The most suitable areas for mango growth include Xuwen, Leizhou, Dianbai, Wuchuan and other places in the southern part of Leizhou Peninsula. Among them, the planting areas of "Qindou mango" in Leizhou are mainly distributed in Qindou, Wushi, Beihe and Yingli, with a planting area of 52,000 mu and an annual output value of about 490 million yuan.

Regional brand: "Qin Dou Mango", China geographical indication protection product.

2.3.2.3 Guangxi Youjiang River Valley medium ripe mango predominance industrial belt

The planting area is about 1.54 million mu.

Variety structure: There are more than 30 cultivated mango varieties, the area ratio of early, middle and late maturing varieties is $5:4:1$, the planting area of more than 100,000 mu of varieties are Tainong No. 1, Guremang 82 (GUI 7), Jinhuang mang and Guremang No. 10.

Available: Late June to late August.

Distribution overview: The mango planting area of Baise city reaches 1.33 million mu, and the yield reaches more than 900,000 t, accounting for about 85% of the total in Guangxi. And 80% of them are concentrated in the Youjiang River Valley, which has become the largest continuous mango production area in China.

Regional brand: "Baise Mango", China National Geographic Indication product; In 2019, it was selected into the China Agricultural Brand directory; In 2020, it was included in the first batch of protected geographical indications of China and Europe. The brand value was valued at 17.3 billion yuan.

2.3.2.4 Southwest Yunnan–South Yunnan–Zhongyuan River basin mango predominance industrial belt

The planting area is about 1.37 million mu.

Variety structure: Yunnan mango varieties are diverse, the main varieties are three-year Mango, ivory Mango, Myanmar No. 3, Kate, Shengxin Mango, Mache Mango, Tainong No. 1, Jinhuang, Guifei and so on.

Available time: May to November.

Distribution overview: Yunnan is the province with the longest supply period of fresh mango in China. Now it has formed three advantageous mango industrial belts, namely early, middle and late ripening in Nujiang River basin, Honghe River basin and Jinsha River basin. There are 71 counties and cities in Yunnan Province that grow mango. Huaping is the largest county with an area of 420,000 mu, followed by Honghe County with an area of 110,000 mu, and Yuanyang, Hekou, Maitre, Xinping, Longyang, Yongde and other places with an area of more than 100,000 mu.

Regional brand: "Huaping Mango", China's National Geographic Indication product, has won the national and provincial characteristic agricultural products advantage area, national Famous and excellent fruit regional public brand, "Yunnan Top Ten Famous" fruit and other titles.

2.3.2.5 Jinsha River Dry – hot Valley late mango predominance industrial belt

The planting area is more than 1 million mu.

Variety structure: Mainly late-ripening mango, early, middle and late-ripening match. Among them, Kaite, Jilu, Repin 10 and other Panzhihua late mango accounts for about 80% of the annual output.

Listing time: from June to December, centralized listing time is from August to November (Figure 2-2).

Distribution overview: Panzhihua mango planting area is 1.03 million mu, with the yield of 545,000 t. Panzhihua has a unique advantage in growing mangoes, which are more mature here than in other producing areas such as Hainan and Guangxi. Large-scale mango planting bases are concentrated in Renhe, Miyi and Yanbian agricultural counties (districts).

Regional brand: "Panzhihua Mango", a geographical indication of Chinese agricultural products, was selected into the second batch of protection list of Sino-European Geographical Indications in 2020.

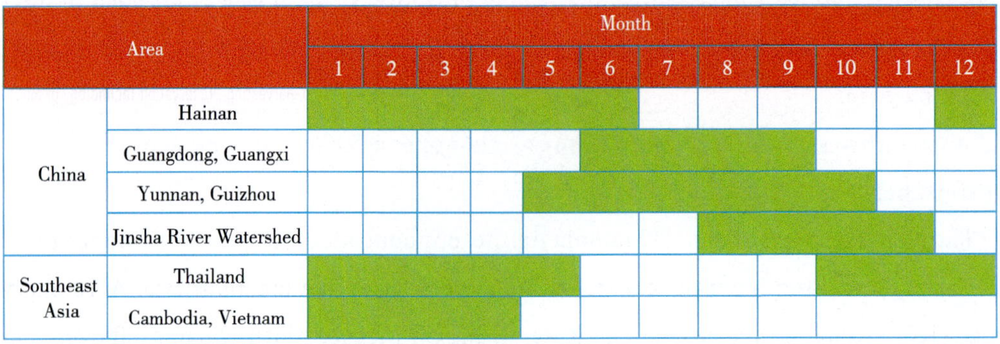

Area		Month											
		1	2	3	4	5	6	7	8	9	10	11	12
China	Hainan	■	■	■	■	■							■
	Guangdong, Guangxi					■	■	■	■				
	Yunnan, Guizhou						■	■	■	■	■	■	
	Jinsha River Watershed								■	■	■	■	
Southeast Asia	Thailand	■	■	■	■	■	■					■	■
	Cambodia, Vietnam			■	■	■							

Figure 2-2 Mango market time of main mango producing areas

In recent years, mango producing areas in China show the characteristics of "early fruit is getting earlier and late fruit is getting later ". The original sales seasons of the two producing areas are staggered, and the overlap degree is getting higher and higher. In a certain market range, mango market is under great pressure of competition, and the ability to resist risks is low, so the price of mango will also be affected.

The Planting Management Techniques

With rich nutritional and medicinal value, The mango is popular with the public and its market prospect is broad. The mango industry is also a characteristic industry in tropical and subtropical areas. Quality management is the key to achieve high yield, high quality and efficient production.

3.1 Mango seedlings and their propagation techniques

The propagation methods of mango can be divided into two kinds: sexual and asexual. Sexual reproduction means to sow seeds and reproduce, and the seedlings are commonly known as "Baosheng seedlings". Asexual propagation, including grafting, air layering and cuttage, is commonly used by grafting.

3.1.1 Sexual reproduction

3.1.1.1 Selection of nursery

You are advised to choose a flat, sunny, sheltered flat or gentle slope. It is better to choose the sandy loam with deep soil layer, good drainage and rich organic matters. You should pay attention to the drainage when you prepare the filed, because the seedlings are afraid of water. If the water soak them for several days, it is easy to lead to stunted growth or death of seedlings. According to the soil fertility, organic fertilizer and superphosphate should be applied as base fertilizer.

3.1.1.2 Pregermination and sowing

It is best to select seeds with full cores. You can select fresh seeds of about 20 d after fresh eating, and the residual meat should be washed and dried in the shade. It is necessary to peel off the shell before sowing, because mango seeds have a hard outer seed coat (shell), which will affect germination. Removing the outer seed coat and covering with soil for 2 cm can effectively improve the germination rate (Figure 3-1).

The seeds will lose their germinating ability after being exposed to the wind and sun for about 7 d. If the seeds are washed and dried a little and stored in semi-humid sandy soil for germinating, they can be preserved for about 30 d. When it is time to sow, seed kernel should be upright and triangular transplanting method should be adopted. Seeds' umbilicus should face down, not plant flat, otherwise it will cause irregular germination. The seeds germinate about 7 d after planting. The planting row spacing should be 10 cm×15 cm, and the seed should be covered with about 2 cm soil. After seeding, you should water every 1–2 d to keep the seedbed moist. When the seed sprouts to the second shoot, it can be applied topdressing once or watered once with 2% urea water solution, and the seedlings can be used as the standard grafting stock after 6 to 9 months.

Figure 3-1 Mango's kernel and embryo

3.1.2 Asexual reproduction

The mango grafting can be used for mass propagation of superior seedlings and renewal of superior varieties in established orchards. Grafting is clonal reproduction, and not easy to change,which can maintain the good characteristics of its mother. As long as a new good variety is bred, you can use the grafting method to conduct mass propagation, such as the Jinhuang,

Guifei and other existing varieties. There are a variety of grafting methods, generally using the cutting method is easier. For the variety not easy to survive like the Tainong No.1, also use the inarching method.

3.1.2.1　Grafting time

In different grafting periods, the graft rate was the highest in March, followed by September, and the worst in June and December. March marks the arrival of spring, and the temperature rises and the climate is mild. It is also the beginning of mango's germination and growth. The plants become stronger after grafting, so the grafting survival rate is high. Most nurseries and fruit farmers choose grafting at this time. In June, due to the high temperature and humidity, frequent rainfall, many diseases and pests and heavy moisture, rootstock and scion interface is easy to mold if you bag the seedlings, which would lead to the result that the grafting healing tissue can not produce or not heal after production. Therefore the rainy season is not suitable for grafting. If the grafting do not meet the rainy season, the survival rate is similar to that of March. September is the crisp autumn season. It is in the end of the rainy season for the southern region. At this time, it is a good grafting period. The autumn shoots of mango plants extract at this time, but the grafting rate is still slightly lower than that in March. December is the beginning of winter. The average temperature will drop below 18 ℃, and most plants are not easy to graft due to slow growth. To sum up, the ideal grafting period is about from late February to May before the rainy season, and March is the best grafting period. If you miss this time, you must wait until the end of the rainy season, and then graft in September to October. The operation should stop before the first cold current.

3.1.2.2　Stock selection

The excellent rootstock should have following advantages: strong resistance to pests and diseases, economical and convenient seed source, fast propagation and growth, strong grafting affinity and ability to adapt to the local climate environment. Its grafted fruits quality will not deteriorate and yield will not reduce.

Grafting stock with 1–2 years old seedling stem diameter of 1 cm is the best, because of its rapid development and high grafting survival rate. If sown in July, well-grown seedlings will have a growth period of about 250 d by March of the following year, and can be used as stock.

3.1.2.3　Scion clipping

Generally, scions should be selected from robust plants with vigorous growth and no pests and diseases. It is better to select mature or semi-mature upright branches in the current year and the upside-down or transverse branches are not suitable. The maturity of the scion must be matched with the rootstock. If the 1 to 2 years old rootstock is used for grafting, you should choose more tender or semi-mature branches. If the rootstock is perennial, scions should be cut from mature branches, and not from the branches of the plant which is still in the juvenile stage, the branches of the plant in flowering or branches of within 30 d after fruiting or harvesting. That is because the scion clipped is not easy to survive when the plant is in reproductive growth.

The best time to pick stock is in the morning. After the scion is harvested, the Tainong No.1 for example must be grafted within 1 hour. The graft survival rate is high if there is latex outflow from the scion, and low otherwise. After harvesting, the leaves should be cut off immediately, leaving a 0.5 cm petiole to avoid injury to latent buds, and grafting should be done immediately. If transportation is required, plastic bags should be used to wrap and store them in the lower layer of the refrigerator. The storage time is about 5 d. The longer the storage time, the lower the viability.

The varieties with low graft survival rate can be treated with accelerating germination or leaf clipping to improve the survival rate. 7 d before grafting, you can cut off the top of the branches intended to be scions. The branches have the advantages of terminal buds, and the latent buds at the top will gradually protrude after about 7 d. At this time, cutting off the branches for scions will improve the graft viability. Leaf clipping treatment is to cut off the leaves of the branches providing scions 7 d before grafting, leaving only about 1 cm petiole. This treatment can also improve the grafting rate. However, if the branches spray with Bordeaux liquid and are cut to be used as scions, grafting is not easy to succeed.

3.1.2.4 Grafting method

Before grafting, you should prepare following things, such as scion, stock, branch scissors, saw, cutter, transparent plastic bag, old newspaper, rope, label and pen (pay attention to grafting date and variety).

In the grafting in the nursery, you can cut the stock 30–40 cm above the ground. After cutting the grafting mouth flat with a cutter, you can cut about 3 cm along the center of xylem and phloem with a cutting knife, and cut the scion into 5–6 cm each section (Figure 3-2). There are two buds left at the top. Then cut the flat side of the scion about 3 cm straight to the xylem. Insert the scion into the stock after about 0.8 cm oblique cutting on the other side of the opposite side. Note that the cambium should be tightened on one side, then tie with plastic rope and cover with transparent plastic bag. Finally, wrapping with old newspaper marks the finish. After grafting, about 21 d at high temperature it can survive, and about 28 d at low temperature you can open the paper bag, and check whether survival. If the grafting is a failure, scion will shrink dry or mold. If the grafting is a success, scion still maintains the original color, and slightly protrudes from the latent buds. Then it will grow new buds and new shoots. If the scion has not grown new shoots when unwrapped, the water in the plastic bag should be wiped off and re-wrapped with newspaper. After 5–7 d you can check again. When the scion grows new shoots, it is advisable to cut or cut the plastic bag, wrap the old newspaper outside, and let it grow.

Adult plants and seedlings have the same grafting method and principle. The only difference is that the scion should select mature branches, and one seedling take a spike. Adult plants can receive 1–3 spikes depending on the size of the branches. Grafting of adult plants is generally for the purpose of updating excellent varieties, so if the original varieties are not grafted, the grafting can be done by sawing to about 60 cm above the ground. If they are grafted and renew the varieties again, you should saw to the original rootstock for grafting. If grafting fails, you can cultivate new shoots on the rootstock and reserve about 3 shoots in different directions. When the shoots mature, they will be attached to the new shoots. Although its future development is a little slower than that on the larger stock, it is easier to succeed .

Figure 3-2　Mango grafting technique

3.1.3　Cultivate rootstock seedlings

3.1.3.1　Selection and preparation of nursery site

You are advised to set up the nursery in the loam or sandy loam close to water source, sheltered from wind and not easy to deposit cold air, with deep soil layer, rich organic matter content and good drainage. There are no mango orchards and old mango trees within 1 km of the surrounding area, so as to reduce the spread of diseases and insects and cultivate seedlings without pests and diseases.

You need to plow and rake the nursery land for three times to keep the soil loose, and the drainage ditch should be prepared. The land with high water table or low-lying lands should be raised. The raised land should be 10 m long, 0.8–1 m wide, 0.2 m high, and the distance between lands is 0.4–0.5 m. You must apply enough base fertilizer and do the shallow ploughing, so that soil fertilizer mixes evenly in order to ensure the needed nutrients for growth of seedlings. 4,500–6,000 kg of manure or compost can be applied for per hectare nursery land.

3.1.3.2 Seeds' collection and processing

The fruit used as stock seed should be selected from a strong growing parent tree, and is fully ripe and fresh. In the production, we can use the seeds left by the cannery processing on the same day, or we can buy a lot of plump fruit, and take out the seeds after the fruits rot. Generally, the local mangoes are suitable for the cultivation of rootstock seedlings because of its high germination rate, neat germination and robust characteristic. The seeds should not be kept for a long time after removal, and the pulp should be washed and dried. Seeds should be sown within 5 d, and the germination rate decreased seriously after 7 d. When transporting seeds over long distances, they can be stored in wet coconut bran, charcoal powder or wet river sand. The seeds used as rootstocks should be the ones with strong affinity with local popularised varieties. Generally, the seeds of large-leaved varieties (including excellent cultivated varieties) should not be used as rootstocks.

3.1.3.3 Peeling the shell to promote budding

The outer woody shell of mango seeds interferes with seed germination, and the germination rate is low if seed cores are sown directly. The bending and deformed proportion of seedlings are large. The method of peeling the shell to promote germination should be adopted to improve the yield and cultivate healthy rootstock seedlings. Practice has proved that the germination rate of shelled seeds is more than 90%, while that of unshelled seeds is only 40%–60% or lower.

3.1.3.4 Separate bed transplantation

The seedbed can be double-row or four-row. The two-row spacing is 20–30 cm, plant spacing 20 cm and bed surface width of 50 cm. The four-row spacing is 20–30 cm, plant spacing 20 cm, and bed surface width of 1 m. A bamboo skewer can be used to carefully dig out the whole root of the seedling and move it to the seedbed with a row spacing of 22–25 cm and a plant spacing of 15–20 cm. Care should be taken not to damage seedling roots during seedling transfer. If the taproot is too long, it can be appropriately cut short, but the root length should not be less than 10 cm. If the seedling is raised in a nutrient bag, the seedling should be moved into the bag containing nutrient soil [1/3 fine sand, 1/3 soil, 1/3 organic fertilizer after maturation and sterilization is mixed and bagged. The bag size should be (22–24) cm×30 cm]. The seedling is moved into the bag for 1 seedling per bag, and you can irrigate the soil with appropriate amount of water.

3.1.3.5　Stock seedling management

Water drenching and fertilization: After transplantation, drench the seedling promptly once a day to keep the seedbed moist. In order to ensure the nutrients required for the growth of seedlings, topdressing can start when the young seedlings sprout new shoots. It can be applied with 1∶5-1∶4 fecal water or 0.5%–1% ammonium sulfate or urea aqueous solution. Do the topdressing once for every 1–2 new shoots.

Shading: When transferring seedlings, you are advised to choose cloudy days. The seedlings growing for no more than 3 months are young and tender. Shade shed should be set up with 50%–60% black shading net above the nursery to protect the seedlings.

Disease control and prevention: in seedling stage, the main diseases include anthrax, leaf spot, and carbendazim and Bordeaux liquid can be used for control and prevention.

3.1.4　Cultivation of grafting seedlings

3.1.4.1　Grafting period

Generally March to May or September to October each year is the most suitable grafting period in which the temperature is stable and humidity is not high. Seedlings is in the before and later period of peak growth season, and the survival rate is high.

3.1.4.2　Scion selection and storage

Scions must be selected from the scion nursery or the mother tree with robust growth, no disease and insects and pure varieties. Healthy, full and 1 to 2 year old branches with many leaves and full eyes should be selected as scions. Cut off the leaves of the scion and pay attention to not hurt buds and graft in time. If you need to store and transport for more than 3 d, you can store with wet coconut bran, or put them directly into the polyethylene film bag, and then box in moist environment.

3.1.4.3　Grafting method

(1) Cutting and connecting method

Its advantage is that it can use the younger scions. It is not affected by phenology and the difficulty of peeling. As long as the temperature conditions are favorable, grafting can be done at any time, and the germination and seedling growth after survival are fast. The detailed operations are as follows.

① Cutting stock: Cut the stock at about 25 cm from the ground. The cutting mouth should be slightly inclined to a straight side, and then cut a incision of about 2 cm at the lower part of the oblique section. The depth should be appropriate to cut off a little xylem, and then cut off 2/3 of the cortex with a little xylem. ② Cutting scion: Use a branch with similar thickness to the stock (small rather than large), select and cut 1 to 2 buds as a section of scion, cut all the mouth from top to bottom on one side. The length is the same as the cut on the stock, and the depth should be appropriate to cut off a little xylem. Then you need to cut the back end of the cut into a 45° oblique. (3) Scion placement and binding: insert the lower end of the scion into the cutting position of the stock, so that the scion and the cortex of the cutting of the stock are matched. Then tie the anastomotic part tightly with ultra-thin film, and then wrap the scion tightly. Without untying, the buds can penetrate the film out. During operation, the knife should be sharp and your actions should be fast. The cambium should be kept complete and clean. the link should be accurate and the binding should be tight.

(2) Bud grafting

Its advantages are simple operation and easy to learn. It also has the advantages of economical scion, fast interface healing and convenient patching, but the rootstock and scion must be easy to peel before budding. The detailed operations are as follows.

① Bud grafting: Select the branch segment with smooth epidermis and no leaves 20 cm above the ground of the stock and carve a rectangular slit of about 1 cm wide and 2.5 cm long with the tip of the knife. The long side should be parallel to the direction of the trunk and reach the cambium. After the bud opening, the cortex will be peeled from the top to the bottom and most of it will be removed, leaving a small part of the end to undertake bud grafting. ② Bud cutting: It is appropriate to select axillary buds or dense jointed buds with full bud eyes, and cut 3–4 cm long and 1.2–1.5 cm wide buds above or below the bud eyes at a position of more than 1 cm. Trim both sides of the buds according to the width of the stock interface, and carefully separate the cortex and xylem. You need to pay attention not to damage the cambium, and finally cut the bud into a rectangle slightly smaller than the interface. ③ Placement and binding of buds: Place the peeled buds in the middle of the bud joint in a consequent and straight direction, and then bind them evenly with elastic plastic film tape until they are completely sealed (Figure 3-3).

Figure 3-3　Mango seedling grafting

3.1.4.4　Grafting management

(1) Unbind and repair

If you use branches grafting method, let the bud out after it sprout through the thin film. when the bud sprouts 1 shoot, then you can unbind it. When untying, you need to use a blade to cut the plastic band on the back side of the interface. The incision should not be close to the interface, and do not cut the cortex. After grafting for 3 weeks, use a grafting knife to cut the binding at the interface, and check whether the bud is still alive 5 d later. If the bud remains green, the top of the anvil can be cut off 1 cm above the interface. If the bud turns brown, it indicates that the bud has been necrotic, and the bud should be grafted at the lower part of the interface (Figure 3-4).

(2) Remove rootstock buds

After grafting, it is easy to grow buds in the base of the stock or the cutting mouth. You need to remove many times in time, so as not to competition between them for nutrients and buds, affecting the growth of buds.

(3) Fertilizer and water management and pest control and prevention

After the grafting seedlings are cut, water should be poured in time to promote the budding and heading, and fertilizer and watering should be applied once each time when it sprouts new shoot. The budding tissue of initial sprouting is tender and susceptible to infestation of diseases and pests, so spraying the pesticide should be applied to prevent and control them.

3.1.4.5 Out-planting

(1) Out-planting standard

Nursery stock varieties, quantity and quality should be verified and marked before out-planting, and its plan should be formulated. In the production, the requirements should have pure varieties, good anvil ear interface healing, developed roots, fibrous roots, 80 cm high seedlings, and at least 3 old ripe shoots, no pests and diseases.

(2) Seedling lifting

The mango seedling lifting can be divided into seedling with soil and seedling without soil. Seedling lifting with soil should be carried out on sunny days. Tools such as seedling starter can be used to keep the soil mass with a diameter of 15–18 cm. After seedling lifting , 1/3 of the leaves are cut off, and then straw is used to bandage the seedlings. To start seedlings lifting without soil, water should be filled 1 d before the start of seedlings lifting. Otherwise the roots will be easily damaged during the start of lifting, and the difficulty of digging seedlings will be increased.

Figure 3-4　Mangoes grafting seedlings ready for out-planting

3.1.5 Planting sites selection and requirements of Mango orchards

3.1.5.1 Garden sites selection

Sand loam with good drainage, deep soil layer (more than 2 m), soft and fertile soil and rich organic matter content should be selected for mango orchards. The soil should have favorable ventilation conditions and no water accumulation. The groundwater level should be ensured to be less than 1 m (Figure 3-5).

3.1.5.2 Orchards reclamation, planning and planting preparation

The reclamation is a basic work in orchard planning, including cutting, clearing, ploughing, leveling the land, opening ditches or ladder openings, excavating planting holes or opening planting ditches according to the baseline, etc.

Dig the planting hole 1 to 2 months before planting, and carry out the hole soil back ditch. First put the topsoil on the bottom layer, then put together weeds, branches and leaves and lime of 0.3 kg, and mix them. Then put back a layer of topsoil, and put into the pond mud, garbage soil fertilizer of 30 kg and 0.2 kg of phosphate fertilizer. Then cover a layer of new soil filled to the ground 20 cm. Put in the mixed pile after retting the livestock manure of 10–15 kg, calcium magnesium phosphate fertilizer of 1–1.5 kg and compound fertilizer of 0.2 kg. Then add back to the soil and mix until it is 15 cm above the ground, and cover the soil 10–15 cm to make the top diameter of about 60 cm soil disc. After 20 d the soil disc is fully sunk, and it should be higher than the ground to prevent the planting hole from sinking, burying seedling root neck and water accumulation, which is not conducive to growth. After the above work is completed, planting can begin.

3.1.5.3 Planting specifications

(1) Density and time of field planting

The most suitable planting season for mango is generally from March to May, followed by September to November. The row spacing of planting plants can be determined according to the variety characteristics, soil fertility, management level and mechanization degree of the garden. The general plant spacing is 3.5 m and row spacing is 4.5 m. About 42 plants is planted per mu. Tainong, Aiwen and Kate trees have potential medium, short trees' body and small crown. They can be properly densely planted.

(2) Method of field planting

Before field planting, the first is to disinfect. The stone sulfur mixture of 3–5 degrees can be used to spray before planting. At the same time you need to carry out classification and pruning.

You can start field planting only after 15 d of applying base fertilizer. When planting, you need to first dig a V-shaped hole in the vegetative soil layer with a depth of 30–40 cm (depending on the length of the seedling root). You can put the seedlings down when the root neck of the bud is 5 cm higher than the ground. For bare-root seedlings, the root system should be naturally extended in layers.Then cover the topsoil in layers and gently compact them. Then cover nutrient soil and compact with feet. If you use a nutrient bag of seedlings, you can tear the bag and take out, and then directly cover the nutrient soil and gently compact it. Then cover with a layer of 10–15 cm new soil to form a nest of soil 10–15 cm above the ground. After field planting, you can immediately begin to earth up after water infiltration to prevent water evaporation and seedling shaking, and then cover with dead leaves on the nest mound to conserve water.

Figure 3-5 Mango standardized planting orchards

3.2 Mango cultivation and management techniques

3.2.1 Tree management

3.2.1.1 Shaping and trimming

Shaping and trimming is to help mango trees to have a good tree structure, make full use of space and sunlight, and regulate the the rhythm of growth and development for easy management.

The method and intensity of shaping and trimming should be determined according to the characteristics of different varieties, planting density and water and fertilizer management and other factors. Generally, the tree body's height should be controlled at 2.5–3 m. The structure of the tree body should be clearly layered, make full use of space and sunlight, and have permeability. Shaping and trimming must be closely coordinated with water and fertilizer management to achieve the desired results.

3.2.1.2　Shaping technology

The structure of mango trees: The above-ground part of mango tree includes two parts: trunk and crown. The crown is composed of central trunk, main branch, lateral branch and branch group, among which the central trunk, main branch and lateral branch constitute the skeleton of the crown, collectively known as backbone branch.

Trunk: The stem from the ground rhizome to the first branch is called the main stem. The height of fixed stem of the general varieties is 60–80 cm, and the height of fixed stem of the varieties with more branches, such as Zihua mang and Yuexi No. 1, can be as low as 50 cm.

Central trunk: As the trunk above the main stem, the central trunk is one of the bearing branch groups. Its retained number is generally related to the variety. The varieties with more branches generally reserve 1, and the varieties with less branches reserve 2.

Main branch: The backbone branch born on the central trunk is called the main branch. It is the main skeleton of the crown of the fruit tree and the growing place of the side branches. This kind of tree has large shape, more main branches, small tree shape and fewer main branches. Every mango tree has 5–7 main branches and no more than 3 in the same layer.

Lateral branches: Born on the main branch, the lateral branches are the main part of the leaves growing, flowering and bearing. They are also one of the skeletons of the crown of the fruit tree and the growing places of all kinds of branches. When shaping, you need to preserve as much as possible. In the first bearing year, most of the lateral branches of mango belong to the bearing mother branch, and the new branches that germinate after fruit harvest become the bearing mother branch of the second year. Generally, 30–40 branches are kept in the whole tree.

3.2.1.3　Trimming method

Short cut: That is to cut off a part of the shoot. It can increase branches, promote shoot growth and rejuvenation, change the apical position between different shoots, so as to change the apical dominant position and regulate the balance of main shoots.

Thinning: That is to thin the branches from the base. Its function is to reduce branching, facilitate ventilation and light in the tree crown, and promote flower bud differentiation and fruiting.

Plucking the heart: Plucking the heart refers to plucking the most tender part of the top of the new shoot by hand or cutting it off with scissors when it grows to a certain extent. It can promote bud enrichment, flower bud formation and early fruiting. and improve fruit setting rate. In shaping young mango trees, heart plucking is often used to promote branching.

Chu meng: It is meant to remove buds.

Ring stripping: That refers to the stripping of a ring of phloem from a branch. In mango production, ring stripping can be used to promote the differentiation of flower buds and facilitate the formation of flower buds. The time and width of the ring stripping should be appropriate. Generally, the width of the ring stripping should be 1/5–1/3 of the diameter of the stripped branch. Otherwise, the expected purpose cannot be achieved.

In addition, bending branches, twisting branches, cutting, root breaking, leaf removal and other trimming measures can be used when needed.

3.2.1.4 Cultivation of high-yielding trees

Different varieties (or types) have different tree shapes. The current commercial planting mango trees mainly have evacuation stratified shape, natural round head shape, natural fan shape. Those trees should be treated differently in trimming.

1) Pruning of fruiting trees

(1) Pruning after picking fruits

The largest pruning of a year is carried out in August-September after harvest, and the aim is to adjust the number and angle of permanent backbone branches in the crown for an even distribution. The main method is to truncate the main branches, and the over dense branches and too many main branches should be cut off appropriately. The cross branches and overlapping branches between and within the crown should be retracted. The pendent branches and the branches with insects should be cut off.

(2) Pruning during the growing period

Pruning in the growing period includes pruning in the fall shoot, spring and summer shoot, etc. The methods of buds picking, thinning shoot and truncating are mainly adopted. Pruning of autumn shoots: 1–3 autumn shoots extracted after fruiting should be retained according to their spatial position, and the rest should be removed. The remaining shoots should be cut off when they reach 18–20 cm, prompting the second autumn shoot. The last autumn shoot should be retained with 18–20 cm long and medium strong branches as the parent branches, and the rest should be removed. Pruning of spring and summer shoots: In off-year with insufficient flowering, a large number of spring and summer shoots will produce flourishing growth, and the new shoots will compete for nutrients severely, leading to weak inflorescences and low fruits setting rate. Therefore, trees with less than 50% flowering capacity should be dispersed and remove buds according to the situation, so as to ensure sufficient nutrients for promoting the normal inflorescences. This pruning is carried out after the spring shoot is pulled out, i.e. after the second physiological drop of fruit (when the fruit has the size of an egg). The purpose of this pruning is to cultivate the fruitless spring shoot into the next year's fruiting branch, so that a large amount of nutrients can fully supply the development of fruit.

(3) Culture fruiting parent branches

Mango trees usually have the top flower bud, and the final shoot can become the fruiting mother branch for flowering and fruiting as long as the shoot gets ripe. In most of the mango producing areas in China, there are few spring shoots and summer shoots of high-yielding trees, so the autumn shoots or early winter shoots extracted after fruit harvesting are generally used as the parent shoots. It is the key for cultivating good fruiting parent shoots to induce the timely stop of growth in autumn shoots or early winter shoots (Figure 3-6).

Figure 3-6　high-yielding tree cultivation

2) Old trees renewal

Canopy renewal should be done before spring budding. The wound should be disinfected with 1% copper sulfate and coated with protective agent such as wax in order to prevent disease, reduce evaporation, promote healing and growth of the cut tip. When the main branches renew, the crown backbone branches should be sprayed with lime milk to avoid sunburn. Attention should be paid to windproof and pruning when fresh branches sprout, and those that sprout at the lower part of the trunk should be removed in time.

3.2.2 Soil fertilizer management

3.2.2.1 Management of soil in the vertical projection of the tree crown

After planting, attention should be paid to the management of soil in the vertical projection of the tree crown, so as to moisturize and prevent rats. Generally it should be covered with straw to maintain a certain humidity.

3.2.2.2 Orchard grass and mulch

For mango orchards newly planted or planted for 1 to 2 years, some grass seeds of legume and gramineae, such as stylosanthes, false peanuts, creeping beans, no spines mimosa, tropical alfalfa, etc, can be sown between the tree lines to cover the soil in the vertical projection of the tree crown. After the cover crops grow up, they should be hoed at flowering or the beginning of flowering, and then clear tillage and cover the soil in the vertical projection of the tree crown, which can eradicate weeds, regulate soil temperature, conserve soil moisture, improve soil structure, and promote the microbial activity of the root system. The appropriate cover thickness is 15–25 cm. In Leizhou Peninsula and southwest of Hainan Province, for the orchards with uneven distribution of rainwater and obvious drought in winter and spring, Those actions have an obvious effect of water preservation and cooling. In addition, the orchard interplanting peanut and soybean should use straw mulch, which can bring good economic and ecological benefits.

3.2.2.3 Soil modification by expanding holes

Except rainy days, high temperature and dry weather and blooming period, soil modification can be carried out in most of the year, but it is suitable in autumn and winter.

Method of soil expansion: the shape and position of the soil pit depends on the shape of the planting hole. If the moat is planted, the soil on one side of the moat in turn every year should be modified. If the planting hole is round, two 1/4 arc-shaped soil modification should be done between plants or rows in turn annually. If the planting hole is square or rectangular, you can dig a rectangular hole between the plants or rows each year in turn, and carry out " 井 " shaped hole expansion. The depth of the hole expansion is about 40 cm and the width is at least 60 cm. The length of the hole should gradually increase with the age of the tree.

3.2.3 Nutrition and fertilization management

3.2.3.1 Nutritional characteristics of mango

1) Nutrient requirements of mango trees

Generally, for the trees that yield 1,061 kg fresh fruit per mu need to absorb 104 kg nitrogen, 27.5 kg phosphorus pentoxide, 119 kg potassium oxide, 88 kg calcium, 47 kg magnesium, 871 g manganese, 174 g boron, 375 g zinc, 435 g copper, 976 g iron from the soil. The nutrient absorption ratio is nitrogen : phosphorus pentoxide : Potassium oxide : Calcium : magnesium =1 : 0.26 : 1.14 : 0.85 : 0.45. In our country, every harvest of 18,668 kg/hm^2 of Jyhua mangoes, fruit nutrient uptake is nitrogen 22.4 kg/hm^2, phosphorus pentoxide 9.0 kg/hm^2, potassium oxide 44.7 kg/hm^2, calcium 3.2 kg/hm^2, magnesium 3.0 kg/hm^2, sulfur 2.3 kg/hm^2. Nutrient absorption ratio is nitrogen : phosphorus pentoxide : potassium hydroxide : calcium : magnesium : sulfur = 1 : 0.40 : 2.00 : 0.14 : 0.13 : 0.10. For each 1,000 kg of deshehari, the nutrients taken away by fresh fruit (average) are nitrogen 693 g, phosphorus 231 g, potassium 1,575 g, calcium 225 g, magnesium 212 g, and the ratio is nitrogen : phosphorus : potassium : calcium : magnesium =1 : 0.33 : 2.27 : 0.32 : 0.31.

2) Changes of mineral nutrients in different growth stages

Generally, the contents of nitrogen, phosphorus, potassium, calcium, magnesium, zinc and boron in mango leaves increased gradually with the growth and aging of autumn shoots, and reached the highest value when autumn shoots mature. When mango trees are in the blooming stage, the leaf nutrients transfer to the flower spike, and the leaf nutrients decrease significantly for more than one month. The consumption of nitrogen, phosphorus, boron, zinc, manganese and iron become more than before. Leaf nutrient decreased gradually in young fruiting stage. At the second physiological fruiting stage (the size of glass beads), the contents of nitrogen, phosphorus and boron increased significantly, but the increase of potassium, calcium and magnesium was relatively small. At the stage of rapid fruiting expansion, the concentration of nitrogen, phosphorus and boron decreased, while the concentration of potassium, calcium and magnesium increased significantly to the highest value. After ripening, mineral nutrient content decreased obviously.

3) Nutritional diagnosis, fertilization response and symptoms of deficiency and excess of main mineral nutrients in mango trees and their corrections

(1) Nitrogen (N)

Nitrogen is the main nutrient element that controls plant growth. It is found in a variety of plant compounds and has the highest concentration in the most active part of plant growth. Young leaves, flowers and root tips all use nitrogen. In orchards, the most likely deficiency is nitrogen. When nitrogen deficiency occurs, the leaves are yellow, and the symptoms first appear in the old leaves, and then spread to the top leaves. As time goes by, the tips and margins of the yellow leaves appear necrotic spots, and the deciduous leaves are accompanied by serious physiological falling fruit, thin branches, small flower buds, small fruits with rich color. Chronic nitrogen deficiency will reduce fruit stress resistance. In general, the optimal annual nitrogen application rate is 400 g/ plant for sandy soil orchards and 600 g/ plant for clay soil orchards.

(2) Phosphorus (P)

Phosphorus is an important nutrient for the beginning of flowering, fruit setting and root development. Phosphorus moves slowly in the soil, so it has a better effect when applied before rain. The symptoms of phosphorus deficiency first appear on the old leaves with necrotic brown spots between the veins, and finally cover the whole leaves, and the leaves with phosphorus deficiency fall off early. Severe P deficiency will lead to a series of consequences such as delayed tree body growth, less branches, small leaves, poor flower buds differentiation, late fruit ripening, and the decreasing yield. In China, 150 g P pentoxide per plant is the appropriate amount of phosphorus applied in a fruit-bearing tree year. When phosphorus is excessive, the absorption of nitrogen, potassium, iron, zinc and copper in the tree body would be affected, and the germination of winter shoots and the growth of many summer shoots occurs, which has an adverse effect on fruit setting.

(3) Potassium (K)

Mangoes have a high demand for potassium, which is essential for the growth of autumn shoots, fruit development and fruit quality. The leaves of the potassium-deficient plants are small and thin. Irregular yellow spots appear on the old leaves at first, and then the dead spots spread along the leaf margin between the veins. The early necrosis of the potassium-deficient plants is limited to the leaf margin, and in severe cases, the dead leaves can remain on the tree for several months with poor coloring. For a fruit-bearing tree, the best potassium fertilizer amount was 500~600 g/ plant in sandy soil and 600~700 g/ plant in clay soil. Excessive potassium application easily leads to nutrient imbalance in the tree body, and reduces the content of calcium and magnesium in leaves, which would also reduce the yield.

(4) Calcium (Ca)

The function of calcium is to regulate mango growth environment, improve mango disease resistance, reduce mango physiological barriers. In the case of calcium deficiency, mango leaves were yellow-green, and the top leaves turn yellowed first. In the case of severe calcium deficiency, except the small area near the leaf tip and the leaf base, the old leaves are all damaged with brown injuries along the leaf margin, accompanied with curled and easy-to-break leaves, crumpled edges, dried top buds, and the atrophic flowers. The root system would be damaged and protruded, and the new root is short and thick, curved, and the tip soon turns brown and dies. Lack of calcium in flowering and young fruit will result in poor pollination, less fruiting and serious falling fruit. The fruit is calcium-deficient at maturity, with the appearance of softening and browning of the tip of the fruit, especially in the pulp of the fruit apex and abdomen, which is called soft rhinosis. Applying lime on acidic soil and spraying 2% calcium nitrate or calcium chloride outside the roots can be used as a response to calcium deficiency. In addition, calcium leaching after harvest can also have some effect. For a fruit-bearing tree, the annual application amount of calcium fertilizer is 0.95–1.26 kg quick lime/plant for clay soil and 0.62–0.87 kg quick lime/plant for sandy soil.

(5) Magnesium (Mg)

Magnesium is an important element in photosynthesis of plant chlorophyll, which has an important effect on the growth of autumn shoots, leaf photosynthesis and fruit setting. When magnesium is deficient, plant growth will be inhibited accompanies with shorter and slightly wider leaves, necrotic tips and margins of mature leaves, dark green main veins, light yellow mesophyll between the dark green area and necrotic area, green veins, early leaf shedding and delayed fruit ripening. Magnesium deficiency begins in old leaves and gradually spreads to young parts. When magnesium is deficient, excessive application of potassium fertilizer should be avoided, and 0.5% magnesium sulfate can be used for foliar spraying in the vigorous growth season. The annual application amount of magnesium fertilizer in mango bearing trees is generally about 1 kg magnesium sulfate per plant.

(6) Sulfur (S)

Sulfur is essential for the formation of chlorophyll and is essential for building proteins. The symptoms of sulfur deficiency appear late and slowly. The symptoms are necrosis along the leaf margin as soon as the leaves turn old and ripe. Within 15–20 d, the whole leaves turn brown and gray, and get brittle and easy to fall off early. Typical symptoms of sulfur deficiency are rare. For a bearing tree, the annual application of sulfur fertilizer should be 40–80 g/plant.

(7) Trace elements

It mainly contains boron, zinc, iron, copper, molybdenum, manganese and so on, which are necessary for the growth and development of mango. Boron and zinc are the main trace elements with low nutrient levels in mango producing areas. Boron-deficient young shoots internode becomes short, and the terminal bud is easy to die. The leaves are thick and brittle, and the tree can not bear fruits. Zinc deficiency results in lobular disease, with clustered lobules occurring in the young and tender part of the top, intervein-vein loss of green, early leaf setting and small fruit in the lower branches. Spraying 0.3% zinc sulfate solution at flowering stage and 0.5% borax solution after fruit setting, once every 20 g, twice in succession, can obviously relieve the symptoms of zinc and boron deficiency.

3.2.3.2 Fertilization amount, fertilization period and methods

1) Fertilization for young trees

For 1 to 3 year old saplings, in order to promote the rapid growth of young mango trees and form developed root groups as soon as possible, fertilizer application should be combined with shoot growth phenology. After planting, it is more reasonable to apply fertilizer once before budding and once when the shoots turn green in sandy soil, and once before shoot in clay soil. Fertilizer should be mainly available nitrogen and phosphate, and you should pay attention to the cultivation of fertility. To form strong root groups, organic fertilizer can be applied 3 times a year, and lime can be applied 2 times a year. Organic fertilizer should be applied to spring, summer and autumn shoots before germinating, and lime was applied to spring and autumn shoots before germinating, respectively. For details, see Table 3-1. Fertilizer application method: Chemical fertilizer can be applied with water, or open half ring ditch combined with irrigation. Organic fertilizer and lime can be applied with ring ditch, or combined with hole expansion.

Table 3-1　Fertilization amount of young mango trees

单位：kg/（plant·a）

Age (Year)	Amount of nutrients				Amount of fertilizer					
	N	P_2O_5	K_2O	Mg	The urea	Calcium super-phosphate	Potassium chloride	Magnesium sulfate	Organic fertilizer	Lime
1	0.075	0.075	0.06	0.01	0.16	0.56	0.1	0.1	60	1
2–3	0.15–0.2	0.2	0.2	0.02	0.33~0.43	1.5	0.33	0.2	60	1

2) Fertilization of the bearing trees

Mango seedlings began to bear fruit in the fourth year after transplanting. In this period, when fertilizing, you should take into account the different stages of flowering, fruiting and fruit development. According to the annual variation rule of mango tree nutrients, it can be divided into 4 fertilization periods according to the phenology of fruiting trees.

(1) Post-fruiting fertilization

Before the first pruning after harvest, the fruit and shoot pruning take away a lot of nutrients, and the tree body nutrition consumption is large. At this time, it is appropriate to apply fertilizer to supplement, so that the plant's growth can quickly recover. Post-fruiting fertilizer should be applied early, usually 5–10 d before pruning. Fertilizers should mainly include organic fertilizers, nitrogen and phosphorus. In addition, that should be accompanied with clearing the mango orchards with lime.

(2) Fertilization promoting autumn shoots

Under the climate conditions in southern China, autumn shoots usually sprout from August to September. The biomass of autumn shoots in the growing period accounts for 30%–50% of the biomass of the whole growth period. At this time, plants accumulate a lot of photosynthates and mineral nutrients to prepare for the differentiation of flower buds. When the autumn shoot sprouts, you should apply once fertilizer, mainly nitrogen, potash fertilizer and organic fertilizer. In addition, if the pruning is late and the fall shoots' sprouting is slow, urea (1%–2%), calcium nitrate (0.5%–1.0%) and potassium dihydrogen phosphate (0.2%–0.5%) can be applied on the leaf surface to accelerate the growth of fall shoots.

(3) Fertilization promoting flowering

The fertilization period of flowering fertilizer is about half a month before flowering, in January in Hainan, Yunnan and most of the tropical areas, and in February or March in other places. This period belongs to sprouting and flowering season, and the demand for nitrogen and phosphorus is strong. In the early sprouting, you can apply available nitrogen and phosphorus fertilizer, with boron (0.2% borax), zinc (0.2% zinc sulfate) and other trace elements for foliar spraying, in order to provide nutrition, which can improve the quality of flowers, promote flowering and fruiting.

(4) Fertilization promoting fruiting

The fertilization period is at the time when the fruit grows to the size of glass beads (before fruit's expansion). At this time, as the young fruit grows rapidly, they need to compete for nutrients from the leaves, so an additional fertilizer should be applied to supplement nutrients, promote fruit expansion and reduce fruit falling. Fertilizers are mainly phosphorus and potassium, combined with nitrogen fertilizer. In addition, the fruit preservation agent composed of gibberellin (2.5 mg/L) and cytokinin (5 mg/L) can be sprayed twice every 15 d at this time.

3.2.4　Mango orchards clearing and its management

3.2.4.1　Purpose of clearing

The first is to prune branches, cultivate a good canopy and tree shape fit for their own orchards. Second, it is suitable to dwarf the tree body to facilitate future management; The third is to clean up the bacterial residue of the mango orchards, reduce the orchard bacterial base; The Fourth is to promote neat shoots, cultivate healthy shoots for the second half year of the off-season to cultivate good fruit-bearing mother branches (Figure 3-7).

Picture 3-7　Clearing of mango orchards

3.2.4.2　Main suggestions for clearing the orchards after fruit picking

1) Trees nursing

(1) Problems

Soil acidification and compaction: At present, mango trees are mainly planted on the red and yellow sandy soil. Over the years, in order to pursue the yield, the more application of chemical fertilizer and less supplement of organic matter have resulted in the continuous accumulation of acidic substances in the soil, aggravating and accelerating the acidity of the soil. The soil turns barren, and harmful metals are activated. The harmful microorganisms increased, especially parasitic fungi. The soil-borne diseases are serious, and the root growth is inhibited. Mango trees are faced with weak growth, malnutrition, or death in serious case.

Soil nutrient imbalance: the soil has been leached and eroded for a long time, resulting in a large loss of organic matter, insufficient or lack of trace elements such as calcium, magnesium, zinc, boron and molybdenum in the soil and the absence of potassium and silicon. In addition, the fixation of phosphorus results in the enrichment of nitrogen and phosphorus. The imbalance between various elements and the absence of beneficial elements lead to antagonism between soil elements. Finally some fertilizers are there but useless.

Blind application of hormones: At present, a large number of hormones, such as polybulozole and Ethel, are used in the process of shoots control, flowering promotion and fruiting expansion in some mango producing areas. Such behaviors would lead to of the prematurity, unbalanced nutritional supplement, and prominent physiological symptoms, such as hormone deficiency, gum flow in the tree, poor growth, low budding ability, poor shooting quality, weak flowering and fruiting ability, and poor fruit quality.

(2) Proposed measures

Improving soil, roots, leaves, flowers and fruits is the prerequisite for increasing production and income.

Improving soil: soil is fundamental for the normal growth of all crops. Only loose, ventilated and breathable soil is conducive to root growth, fertilizer and water absorption. Too much acid or alkali in the soil will inhibit root growth, cause root dysplasia, root aging, few new root groups, and seriously cause dead roots, poor fertilizer and water absorption, and weak plant growth.

Promoting good roots: Nutrients of the tree are mainly absorbed by the root system through the soil, and foliar fertilizer only plays a supplementary role. If the root system is not growing well, it is difficult for the fertilizer and water to be absorbed, transported and transformed for the growth of the tree, and the nutrition is unbalanced. Although the tree can sprout shoots, it will show difficulty, irregular, thin shoots and short shoot clusters, leading to the corresponding hormone deficiency symptoms, which would have great impact on the cultivation of robust shoots and fruit-bearing parent branches, and then affect the quality of flowers and fruits. At the same time, due to the nutritional imbalance of the tree, the gum flow phenomenon is aggravated and the tree is weakened, which is likely to bring about bacterial keratosis.

Supplementing organic matter: Supplementing organic matter to the soil and improving soil fertility are conducive to the cultivation of soil beneficial microorganisms and keeping the soil loose and not easy to compact. The organic matter also can adjust the pH of the soil, promote and protect roots, and bring about more comprehensive fertilizer and water absorption, laying a good foundation for the cultivation of robust branches and fruit-bearing branches.

2) Cleaning the tree body

In recent years, as mango bacterial keratosis has become more and more serious, even the trees of whole garden will be infected, resulting in the death of the branches, the fruit surface densely covered with disease spots, and serious damage to the fruit surface, affecting the value of fruit products. Bacterial keratosis is also one of the main diseases of mango trees, and the infection rate is very fast. It needs to be controlled by cleaning the tree body and other measures.

When pruning and cleaning the garden after picking fruits, you can throw out the pruned branches outside the garden and burn them. At the same time, you can disinfect the remaining bacteria, such as spraying 750 times Gaoshang or 1,500 times 50% Chunlei Wangtong wettable powder for 2–3 times around the trunk and branches of the tree crown for prevention and control, in order to remove bacterial keratosis, anthrax, red spot disease and other germs, and reduce the base of pathogenic bacteria, preventing large economic losses.

3) Promoting and sprouting shoots

The quality of shoots is directly related to the uniformity of sprouting shoots and the plumpness and robustness of shoots, which will affect the cultivation of fruits-bearing parent branches, flowering of tree body, fruit yield and quality. At the same time, irregular sprouting shoots also bring a lot of inconvenience to management. Therefore, shoots promotion is an important link. If this key point is not done well, a series of management work will be affected. In order to promote good shoots and cultivate healthy shoots, it is necessary to raise trees, change soil, promote roots, supplement organic matter and large and medium elements required by plants. Only by improving both soil and shoots can we cultivate healthy shoot better.

3.2.5　Water management

3.2.5.1　Irrigation

Irrigation methods include ditch irrigation, tree tray irrigation, drip irrigation and so on. Usually, it should be kept as dry as possible 60–90 d before flower bud differentiation. After normal flower bud differentiation, the plants will go to heading → flowering → fruiting → expansion. This period coincides with mango flowering and fruit rapid expansion, and it is also the key period that requires the most water, about two months before fruit → hard core → ripening → harvest, which is the period of low water demand. The soil should be kept dry to enhance the sweetness and quality of the fruits.

3.2.5.2　Drainage

Avoid standing water in the mango gardens. Mango trees is more tolerant to drought but not to water-logging. This is because the roots of mango trees have strong respiration and high oxygen demand. Poor drainage will inhibit root respiration and reduce absorption function. At the same time, when the soil moisture content was high, the flower bud differentiation will be inhibited and the shoot growth is promoted. Long-term water-logging will lead to falling leaves, falling flowers, falling fruits, dead branches, rotten roots, affecting the growth and development of the tree body.

3.2.6 Management of flowering, fruit setting and perinatal regulation

3.2.6.1 Flowering, fruit setting stage management

Flowering and fruit setting stage is a crucial period in mango production management, which directly affects the yield and quality of mangoes. Therefore, developing different management methods for different growth stages is the most effective way to improve mango yield and fruit quality. Flowering and fruit setting stage management have the following aspects.

(1) The adjustment of flowering shoots

About 2 months before flower bud differentiation, you can remove the over-dense, weak shade and diseased insect branches, and leave only 1 to 2 shoots per branch, which can increase the transmittance of the crown, facilitate the concentration of nutrients, form good ventilation and light conditions, promote flower bud differentiation and prevent the occurrence of diseases and insect pests.

(2) Breeding of pollinating insects

Mango pollination mainly depends on flies which can be raised when mango head sprouts. It takes 14–20 d for mango from heading to florets opening, and 14–20 d for breeding flies from laying eggs to adult worms. Therefore, the best pollination effect can be obtained by breeding flies at heading time when the florets open after adult worms. Breeding methods are to place pig

Figure 3-8　Mango flower spikes

offal or smelly fish in order to attract flies to lay eggs and increase the number of flies, promoting mango pollination fruit setting.

(3) Application of trace elements and calcium fertilizer

Mango trees have a great demand for trace elements and calcium from flowering to young fruit stage, so spraying 0.2% zinc sulfate and 0.2% borax solution once 10 d before flowering and after blooming and fading can promote pollination and seeds setting.

(4) Thinning fruits and pruning fruit branches

When the young fruits grow to peanut size, thinning fruits should be carried out to eliminate diseased insects, deformed fruits, in order to improve fruit quality and commercial fruit yield. Generally, only 2–4 fruits are retained per ear, and the central position is the best. In addition, the newly sprouting tender shoots should be cut off during this period.

(5) Promoting coloring

Some varieties with red skins, such as Hongmang No. 6, Aiwen, etc., should remove the weak branches and insect branches that shade the fruits when you prune in the middle fruiting stage, so that the fruits can receive full light and uniform coloring.

(6) Bagging

The size of mango bags varies according to the variety. For the mango varieties with large fruit shape such as golden yellow and ivory, paper bags with length of 30 cm and width of 20 cm can be used. Paper bag material can be white wax paper, black kraft paper or silver kraft paper and non-woven fabric.

Bagging time is generally after the stable fruit setting, namely the end of the second physiological fruit dropping when fruit grows to the size of the egg. Bagging too early will leave more empty bags, and waste manpower and material resources. Bagging too late will lose effect. Bagging time of mango varieties with red skins is different from that of common varieties. For example, the Aiwen, bagging 30–50 d before harvest is more conducive to fruit coloring. Jinhuang mango can have an early bagging after which the fruit surface will be more delicate and smooth, and good fruit powder will be produced. Kate is a late ripening variety. Bagging early will cause the fruit appearance of poor coloring, so you can bag later to help the fruit coloring.

Bagging method: you can spray sterilization and insecticide mixture onto fruits before bagging, or use the bags which have been soaked into the disinfectant solution equipped with sterilization insecticides. Then select normal and disease-free fruits one by one, tie the bag mouth and fruit stem, and leave a leak hole at the bottom of the bag, in order to eliminate standing water in the bags (Figure 3-9).

Figure 3-9 Mango bagging management

3.2.6.2 Perinatal adjustments

(1) Regulation of flowering period

The blooming time of Hainan mango is generally carried out in August to October, which is in the autumn shoot period. Potassium nitrate and paclobutrazol are commonly used for regulation. Polybulozole can be applied in soil or spray. The validity period of soil application is relatively long, and it has an effect within 3 years. However, soil application should be carried out in advance generally when the second tip sprouts out 10–15 cm long. The method of soil application of paclobutrazol is to dig 15 cm deep, 10–15 cm wide ring groove in canopy projection (drip line) or dig two symmetrical 15 cm deep, 10–15 cm wide, 40–50 cm long straight groove 40–50 cm away from tree head on each side, then mix paclobutrazol and appropriate water in a ditch, and cover them with soil. Apply 8–10 g per square meter of canopy projection area. When spraying, you can use potassium nitrate combined with one or two kinds of mixture, such as polybulozole, ethoxazole, netoacetic acid, ethrel, Kongshaoling, Zhaoshaoling, and Aizhuangshu. The application amount should be determined according to the tree crown size and tree potential. Generally, the healthy trees are treated with 15% polybulobulozole 25–40 g per plant, and the weak trees are appropriately reduced.

(2) Improving the fruit setting rate and fruit preservation

In terms of improving the mango setting rate and increasing the yield of fruit by medical measures, 70 mg/L of gibberellin and 30 mg/L of naphthylacetic acid can be sprayed to improve the mango setting rate at flowering stage, and the effect of increasing the yield is obvious. Generally use 0.3% borax or 0.2% boric acid spray, spray once every 10–15 d, and spray after the dew drys in the morning, or spray in the afternoon. You need to pay attention to avoiding spraying under the noon blazing sun.

(3) Use early, middle and late maturing varieties

Adopt early, middle and late maturing varieties to lengthen the market supply period and improve the market competitiveness. At present, mango growing areas are relatively single, basically ripe varieties. The mature period is more concentrated, which is bound to affect the fruit sales treatment. Therefore, for regional planting mangoes, it is recommended to use early, middle and late varieties to plant together, in order to achieve the adjustment of the perinatal period.

(4) Using regional differences to regulate the market

After adopting the perinatation regulation technology, the early fruit in Sanya, Hainan will be on the market from February to April, while the regular fruit in Zhanjiang, Guangdong will not mature until the early middle of June at the earliest, and the fruit in Panzhihua, Sichuan Province will not mature until 30 to 40 d later than that in Guangdong. Therefore, the market regulation can be made by using the differences of climate in different regions and the differences of different varieties in maturity period.

Main Pest Control and Post-harvesting Preservation Technology of Mango

As a major tropical and subtropical fruit, the mango industry has developed rapidly in recent years. At present, the mango planting area in China has exceeded 5 million mu. Although the output of mango per unit area in China has increased significantly, the quality has declined, and pests, as one of the main factors affecting mango quality, are getting worse and worse, which has become an obstacle to the development of the mango industry and seriously threatens the safety of mango production.

4.1 Main pests and control of mango

4.1.1 Mango cut leaf Weevil

4.1.1.1 Disease symptoms

Adults eat the tender leaf epidermis and leaf flesh, gnaw the spots almost round, leaving only the transparent lower epidermis, and the leaves curl up and dry. Female adults lay eggs on tender leaves and bite off near the base of the leaves. The incisions are neatly cut like a knife, causing bald shoots and seriously affecting the tree.

4.1.1.2 Living habits and occurrence characteristics

The annual generation algebra of mango cutoff weevil varies from region to region. There are 9 generations in Hainan, 7 generations in Guangxi, and 3-4 generations in Xishuangbanna, Yunnan. There is no overwintering phenomenon in winter. During the laying and clipping period of adult insects, these clipping leaves will wither rapidly in the hot sun, eventually leading to the death of a large number of eggs and larvae. Soil moisture content has a great impact on the growth and development of pupa. Less than 10% or more than 20% can lead to the death of pupa in the early stage. After eclosion, adult mango cut leaf weevil (Figure 4-1) showed obvious upward, tender and clustering characteristics, and often concentrated in the tender shoots and young leaves of mango. Landing or flying away in fake death when alarmed.

Figure 4-1　Mango cut leaf Weevil

4.1.1.3　Control methods

Agricultural control. ① In the newly opened orchard, it is necessary to avoid the mixing of mango and longan to eliminate or reduce insect sources; ② Orchards mixed with mango and longan can be combined with weeding, fertilization or control of winter shoots to pine the soil and kill part of the pupa and winter larvae in the soil.

Artificial control. During the occurrence of weevil, it is necessary to collect the mango tender leaves that have been bitten to the ground in time, eliminate insect eggs, and reduce the insect mouth of the next generation. From the beginning to the harvest period of mango's second physiological fruiting, attention is often paid to picking up ground fruits and burning abandoned cores.

Pharmaceutical prevention and control. During the feathering period of various generations of adults, master the insect situation and spray at an appropriate time to kill adults. Effective agricultural drugs are: 90% crystalline trichlorfon, or 80% dichlorvos 800–1,000 times liquid, or 20% quick kill Ding or 2.5% enemy kills 2,000–2,500 times liquid, or 80% dichlorvos plus 40% 1,000 times each.

It is necessary to strictly implement the quarantine system and strictly prevent pests from spreading outside the epidemic area with fruits, fruit cores or seedlings. Once the new area is discovered, it must be extinguished in time.

4.1.2　Rhytidodera bowrinii white

4.1.2.1　Disease symptoms

Larvae drill the branches of fruit trees such as mangoes and cashew nuts, usually invade from the young branches, drilling decay branches and trunks, drying the branches and affecting the growth of plants. When the wind blowing, it often causes broken branches or trunks to fold, eventually leading to the death of the whole plant.

4.1.2.2　Living habits and occurrence characteristics

One generation occurs once a year. After hatching, the larvae are drilled into the branches and medullary to eat, and white colloids often accumulate at the end of the tunnel. After the larvae are ripe, they turn pupasalis in the tunnel. After pupa feathers, they stay in the pupa room for a period of time and then go out to eat young branches, buds and tender shoots (Figure 4-2).

Figure 4-2　Longicorn larvae and adults

4.1.2.3　Control methods

Strengthen orchard management. We can strengthen the tree and improve the resistance of the tree. Combined with pruning, cut off the affected branches in time, clean up the orchard, the residues can be destroyed and the source of insects can be reduced.

Manual prevention and control. According to their living habits, wire can be poked into the moth for artificial capture. For serious damage to fruit trees, the crown can be truncated and strengthened.

Chemical control. 2.5% deltamethrin or 20% fenvalerate can be used to plug cavities in the tree trunks.

4.1.3　Thrips

4.1.3.1　Disease symptoms

Thrips is a common pest of mango. This pest uses adults and nymphs to suck the juices of mango's stems, leaves, flowers and fruits, causing the plant to wither. The edges of the affected leaves cannot be curled and ripple-shaped. The leaves become narrow or longitudinally wrinkled. The veins are pale yellowish green, and the leaves have yellow contusion points, which are like flowers and leaves. Finally, the leaves lose luster, stiff, yellow, brittle and fall off. The terminal buds of the new shoots or seedlings are affected, and the clumping phenomenon or the terminal buds show atrophy (Figure 4-3). The stung fruit surface epidermal oil cells rupture, gradually lose water and shrink, and the scar expands with the expansion of the fruit, showing a wooden bolized shape.

4.1.3.2　Living habits and occurrence characteristics

The peak of thistle occurrence in mango was closely related to the phenological period of mango. The number of thistle infestation reached its maximum from the initial flowering stage to the full flowering stage. When the small fruit stage arrived, the number of thistle infestation would decrease significantly. When mangoes are growing, flowering and bearing fruit, the damage is more obvious in warm, dry weather.

Figure 4-3　Thrips and its harms

4.1.3.3 Control methods

We need to do well in pre-flower trimming, remove overly dense branches and leaves, and increase the light transmittance of the canopy. You can also shake the residual flowers every 2–3 d during the flowering period. The purpose is to disrupt the thistle sunscreening site and reduce the density of the insect mouth.

Control can be used alternately with 10% imidacloprid wettable powder 1,500 times liquid and 0.1% abamectin wettable powder 1,500 times liquid. Also, two kinds of agents can be mixed with imidacloprid and abamectin and sprayed with 1–2 times at the bud stage and after the flower has stopped.

4.1.4 Chlumetia transversa

4.1.4.1 Disease symptoms

Chlumetia transversa is known as the drill worm and its larvae often drill into the young tree tops of mango, eat mango flower ear, young leaves and other parts, resulting in tree tops wilt, fallen leaves, seriously affecting the normal growth of plants, resulting in a decline in the amount of results.

4.1.4.2 Living habits and occurrence characteristics

Generally, it breeds for 8 generations a year and overlap in generation. Adults come out day and night, and the phototaxis and chemy are weak. The larvae are generally harmful to the young shoots from May to June and August to October, and February are harmful to the buds and young shoots (Figure 4-4). The larvae have 5 years in total, and the mature larvae spin silk to seal and pupate in the dead wood, dead branches, bark of mango or other insect shells and longan dung.

Figure 4-4   Chlumetia transversa and its harm

4.1.4.3 Control methods

Artificial prevention and control. In the concentrated stage, cocoa-coconut husk (pinebedding) wrapped in plastic can be used to attract the chlumetia transversa. A 10 cm wide agricultural plastic film was cut around the trunk, and a 10 cm long interface was left overlapping. The lower end of the film was tightly tied to the trunk with a packing tape. The upper end of the film was pulled apart to form a trumpet shape and filled with moist coconut bran (pinebedding) to induce the larvae of the chlumetia transversa to pupate inside.

Biological control. Protection and proliferation of parasitic predators; Keep chickens for pest control.

Chemical control. Spraying when mango shoots and flower panicle are 3–4 cm long can protect the tender shoots from this insect. Effective agents are: 40% dimethoate 1,000 times liquid; 50% Daofeng powder 800 times liquid; During the oviposition period, 90% trichlorfon and 20% trichlorvos can be used to kill eggs with 1,000 times liquid, respectively.

4.1.5 Erosomyia mangiferae felt

4.1.5.1 Disease symptoms

The rhopalomyia californica feeds mango tender leaves. The damaged tender leaves appear brown spots after the white spots, the broken leaves are curled. In severe cases, the leaves wither and fall off, resulting in the ends wither (Figure 4-5).

4.1.5.2 Living habits and occurrence characteristics

Mating and breeding in large numbers during warm and humid seasons, adults prefer moist and shaded environments and are afraid of bright light. After being bitten by the larva, the leaf wound is yellowish white tumor, and the leaf shrinks. When humidity is high, especially after rainy day, the wound is susceptible to anthrax infection, which seriously affects the growth of mango shoots.

4.1.5.3 Control methods

pruning and clearing the garden in time is essential for keeping the canopy in the orchard ventilated and transparent. In order to destroy the breeding, reproduction and pupa of leaf gall mosquitoes, we can weed in time and loosen the soil in time. At the same time, we also need to pay attention to reasonable fertilization and scientifically use of water, promote neat, strong new shoots and improve resilience. In areas where gall mosquitoes have occurred, it is necessary to cut off the branches and leaves of diseased insects and shade branches in the garden after fruit picking, and concentrate on the harmless treatment of dead branches and leaves in the garden.

Figure 4-5 Erosomyia mangiferae felt and its harm

4.2 Major diseases of mango and its prevention and control

4.2.1 Anthracnose

Anthracnose is the most common and most harmful fungal disease in mango production. It can cause long-term leaf spots, shoots, deciduous leaves, fallen flowers and fruit rot by infecting mango leaves, tender shoots, flowers and fruits. This disease is easy to occur in large quantities under high humidity. The mango blossom and young fruit stages are more susceptible. Ripe fruits are susceptible to diseases, and the young shoot stage is the most serious (Figure 4-6).

4.2.1.1 Disease symptoms

It is a fungal disease that can occur all year round, especially in the wet and rainy season. There are small brown spots in the early stage of the diseased leaves, surrounded by yellow halo. The spot expands into a circular or irregular shape, dark brown, and several spots fuse to form large spots, causing most of the leaves to die. After the fruit feels sick, it only shows a small needle-like black spot, which lurks until the fruit is ripe. Moreover, its incubation period is highly infectious.

Figure 4-6 Mango anthracnose

4.2.1.2 Control methods

Field hygiene: trimming, removing, burning, selecting disease-resistant varieties (Tainnong No. 1).

Field management: Ventilation and light transmission, weed removal.

Chemical control: It can be sprayed in time according to the growth period and weather conditions of mangoes. The more commonly used fungicides are 1∶100 Bordeaux liquid 500 times liquid, 40% polymycin 200 times liquid, 25% Daisen zinc 400 liquid, 75% Baishiqing 500 times liquid, and 70% methyl thiocillin 1,000–1,500 times liquid have control effects. In addition, 1.5% polyoxymycin 300 times liquid is sprayed every 10 d during the bud period, 2–3 times in a row. During the fruiting period, it can be sprayed once a month. From the sprouting period, it can be sprayed once every 7–10 d, and 2–3 times in a row.

Post-harvest treatment: Harvest the fruit in time after ripe, minimize the wound, and soak and wash the fruit with water after harvesting.

4.2.2 Powdery mildew

4.2.2.1 Disease symptoms

It is a fungal disease, concentrated in places with a dry climate. It mainly affects the flower panicle, leaf and young fruit of mango, and the main occurrence period is from flowering to young fruit, which can cause serious flower and fruit falling, and then affect the fruit setting rate. The obvious feature is that the disease is densely covered with a layer of white powder. At the beginning of infection, there are only small white powdery spots, and then gradually expand into large patches to form a layer of white powdery objects. When the mango flower is infected, its calyx, stem and stem are gradually covered by this white powder, and then turn black until it withers (Figure 4-7); when the mango fruit is infected, the surface of the fruit will be covered with white powder, making it easy for the fruit to fall off easily.

Figure 4-7　Mango powdery mildew

4.2.2.2　Control methods

Physical control. do a good job in field sanitation, and concentrate on removing or burning dead branches and leaves.

Chemical control. Commonly used reagents such as thiamine preparations, powder rust, sterilization, nitrate mite, etc.

Spray timing. The key period is the flowering period. After 5–10 cm of inflorescence, start spraying, and spray once every 10–15 d or spray powder with 320 sieve sulfur powder once during bud extraction, flowering and fruiting period. Reduce the concentration during the flowering period to avoid drug damage.

Agricultural prevention and control. control the excessive application of chemical nitrogen fertilizer, and increase the application of high-quality organic fertilizer and phosphorus potassium fertilizer. Cultivate mango varieties with good resistance, such as autumn mango, No.1 in western Guangdong, Luzon mango, etc.

4.2.3 Downy mildew

4.2.3.1 Disease symptoms

Downy mildew, which is caused by lower fungi and spread by air flow, is easy to occur in damp air and humid environment during the mango flowering period. It infects the mango flowers and produces moldy layers of bloom in white, purple gray, gray brown, black or other colors, resulting in black flowers, rotten flowers, or even no fruit bearing during the flowering phase. The mango trees which are lightly infected will reduced the output while the seriously infected ones will have no harvest at all. Normally, it easily infects the mango flowers when they are too many and crowded together, when they are in poor transmission and ventilation, when there are many clusters of flowers (Figure 4-8).

Figure 4-8　Mango downy mildew

4.2.3.2 Control methods

First of all, the mango trees should be cultivated properly and cleaned completely in the orchards in winter; Secondly, the flowers should be shaken and washed after they entirely fell; Thirdly, when the main inflorescences stretched to 3-5 centimeters long (or the time of inflorescences were lasted from 5-7 d), they should be fertilized and avoided to huddled together; Fourthly, the flowers on the inflorescences should be artificially thinned and shortened (the part of late-maturing areas) to enhance good ventilation and light transmission; Fifthly, the trees should be protected by chemical medicine at the early stage of flowering and flower fading.

4.2.4 Stem end rot

4.2.4.1 Disease symptoms

The disease mainly damages mango fruits: in early stage, the diseased parts (close to stem end) will grows with water-soaking-shape and irregular diseased spots in brown, which turn to crineous till black, and expand to the end of the fruits quickly. At the very beginning, the stem ends are lusterless and in crineous. The boundary between ill and healthy parts is clear. Soon, The diseased expand to the other parts of the fruit and the surfaces turn from crineous to seal brown or purple under the damp and heat condition. At the same time, the sarcocarp softens, the juice flows out with strong sweet honey smell and the whole fruit rots in 3–5 d. There are dense black dots, called as the pycnidium of the disease, appearing on the surface of the fruit (Figure 4-9). The cells of the spore are black and lustrous, which could invade the core from the wounds or the surface holes and lead to skin spots. It could also injure the branches from the holes of cutting, causing the bark withering, splitting and flowing out gum as well.

Figure 4-9　Mango stem end rot

4.2.4.2　Disease characters

The former infection source is from the dry branches, barks and fallen leaves by the main transmission: rainwater. The mature conidium could germinate in the germless water in 4–5 h and the conidium can invade from the wound, cutting or other mechanical injury of stems. The germs hide in infecting before the fruits harvesting so the fruits rot quickly after they ripen.

Generally speaking, it is a kind of disease to hide and invade in the early stage and to appear on the fruits till the fruits ripen and are picked up. So the prevention and controlling should be paid more attention to during the whole growth and harvest period. Anthrax can be used as a reference for early cured medication and one fruit should be cut for twice to reduce the mechanical injury and soaked with prochloraz.

4.2.4.3　Control methods

Keep the orchard clean and reduce the source of primary infection. After the pruning, the dead branches and rotten leaves shall be dealt with properly in time in the orchard. And the branches shall be cut as close as possible to the bifurcation to avoid withering of branches when pruning. The method of "one fruit and two prunes" is adopted during fruit harvesting, which can reduce the speed and probability of pathogen invasion from the fruit stalk. The so-called "two cutting of one fruit" refers to: the first cutting, the length of the reserved fruit handle is cut and reserved about 5 cm at the time of harvesting in the orchard; the second cutting, the length of the reserved fruit handle is reserved about 0.5 cm after harvesting but before processing of the plants. Each time when the fruit handle is cut, it should be placed downward to prevent the latex from polluting the fruit surface and the fruit cutter should be dipped in the agent (75% alcohol). In addition, do not use calcium containing compounds, such as calcium containing foliar fertilizer, to the field before fruit picking. Storing the harvested fruits at 10–13 ℃ can also delay the occurrence of stem rot.

4.2.5 Gummosis

4.2.5.1 Disease symptoms

One of the fungal diseases in the process of mango planting, which mainly damages the trunk, branches and petioles of mango and causes drying, can also damage mango seedlings and fruits (Figure 4-10), and has occurred in many major mango planting areas in China. The essence cause of gummosis is the excessive synthesis of gum polysaccharide when it is in high temperature, high humidity and hidden environment.

4.2.5.2 Control methods

Firstly, improve the cultivation of plants and strengthen the management of orchards. Make the trees stronger and water them reasonably by applying more organic fertilizer, mainly phosphorus and potassium fertilizer to improve plant resistance, to cultivate strong tree vigor and alleviate environmental stress on trees.

Secondly, clean the orchard in time. In combination with plastic pruning, the infected

Figure 4-10　Mango gummosis

branches and shoots (20–30 cm below the diseased part) shall be cut off in time, and they shall be taken away from the orchard to burn, so as to reduce the source of initial infection in the orchard, and the wound shall be disinfected. Bordeaux solution can be used for garden cleaning and disinfection (pay attention to spraying on the back of leaves and the internal branches and shoots).

Thirdly, prevent the injurious insects such as longicorn beetles and spodoptera striata from harming the trees. In order to prevent insect pests and to reduce the wounds, the tree body can be painted in white. In addition, attention should be paid not to create wounds during the cultivation process.

The fourth is the diseased part on the trunk. You can cut the diseased part with a sterilized knife to the healthy part, and then apply copper hydroxide or copper calcium sulfate or difenoconazole to the wound.

4.2.6 Dewy spot

4.2.6.1 Disease symptoms

Dew spot is a disease of mango, which can infect the surface with "dew" like spot mainly by the pathogen of *Cladosporium globosum* and *Cladosporium cladosporium* during the growth of mango. The spot is more obvious when the dew condenses on the fruit surface in the morning.

It is commonly known as "mango dew spot" by mango grower, which has a great impact on the appearance and quality of mango. The disease is related to the thickness, toughness of the fruit surface and tree vigor of the peel. It may become more serious when the fruit is close to maturity (Figure 4-11).

4.2.6.2 Disease characters

The disease is more prone to infect the varieties of mango with rich nutrition on the fruit surface and affect the the trees seriously in the orchards with dense shade, dense branches and leaves without ventilation, especially the old orchards under poor management, inadequate pruning, high

Figure 4-11 Mango dewy spot

temperature and humidity in mountainous areas or low-lying areas with vigorous weed growing. The research shows that the infection of dew spot pathogens are more likely to happen under the conditions with temperature of 28 °C and relative humidity greater than 90%. Therefore, the situation is more serious when the plants are in the soil and air of which humidity are relatively high in continuous rainy and foggy weather and special weather conditions such as "first rain", especially in rainy, hot and humid weather from May to June. The orchards which are repeatedly used by organic emulsifiable concentrates pesticides, especially thifenolone, gibberellin, ethephon, etc., are vulnerable to invasion, and the incidence rate is high. At the beginning, a large number of disease spots will appear on the fruits of the rifled branches in the crown, and then gradually spread from the fruits of the branches below the crown to the top fruits in about 3 d. It is found that if one of the two fruits on the same branch is infected, the other fruit will soon have the same symptoms.

4.2.6.3 Control methods

Firstly, select the disease-resistant varieties. Select disease-resistant varieties and plant disease-free seedlings according to local conditions in combination with the climate characteristics in the mango planting area. Don't mix the different varieties of different phenological periods together and try to plant different mango varieties apart according to regional planning.

Secondly, prune frequently and timely. After mango is picked and before new shoots are sprouted, the tree body should be properly trimmed in sunny days. The cutting wound should be timely coated with callus preservative film to help the wound heal and prevent the invasion of bacteria. At the same time of trimming, remove the free branches, excessive dense branches, cross branches, pest and disease branches and clear the shady mango branches in time. Control the tree crown and carry out the flower and fruit thinning at the same time, so as to preserve the appropriate fruit bearing amount of fruit trees and enhance the ventilation and light permeability of the tree crown in the orchard.

Thirdly, clean the soil in the countryside timely to prevent the vector insects. Do a good job of garden clearing and thoroughly clean up the diseased and dead branches before mango flowering, burn them deeply in the soil together, or spray the quick-decay- agent to accelerate their decay after centralized stacking, so as to reduce the amount of primary infection bacteria. Select the low grass species that are not pathogen hosts to plant to cover mango tree rows with grass. At the same time, pay attention to cut grass regularly to maintain the appropriate height and growth of grass species in rows, so as to maintain the biodiversity and ecological balance of the species in the orchard.

Fourthly, pack the fruits in bags to reduce the probability of disease infecting in time. Generally, mangoes are bagged immediately after the second physiological drop to protect the fruits from disease. When bagging, it shall be ensured that the folding of the bag mouth is good for diversion, and the binding shall be tight to prevent rainwater from entering the bag. In addition, the dead big branches should be sawed off in time, and Bordeaux pulp should be used to seal the sawed wound to reduce the probability of disease transmission.

Fifthly, make the trees stronger, water and apply the fertilizer to them reasonably. Reinforce the management of water and fertilizer, which includes using formulated fertilization, making the balance of tree nutrition, improving its disease resistance, and promoting fruit tree shoot pulling and flowering in order, so as to centralized control the fields. First of all, prepare the drainage and irrigation facilities of the orchard well. Drain the water timely in rainy days to avoid water accumulation increasing the humidity of the orchard. Next, increase the application of organic fertilizer. Generally, dig ditches and bury the organic fertilizer and beneficial bacteria fertilizer into the root of the crown when pruning at the young fruit stage of mango, in order to keep the tree body healthy and improve its stress resistance. Finally, pay attention to the reasonable application of nutrient, especially calcium, boron, magnesium, potassium, etc. Supply calcium and boron fertilizer once on the leaf surface of mango during the flowering stage and from the small fruit to the expansion stage as well, which is conducive to the mango flour, and reduces the probability of water droplets directly contacting the peel; spray oligosaccharin once every 10–20 d during the period from expansion to coloring of fruit, and supply potassium and magnesium at the same time to increase the growth of mango and increase the thickness of inner peel; In addition, the use of foliar fertilizers containing hidden illegal ingredients on mangoes should be avoided.

Sixthly, use the regulators that could adjust the plant growth properly to reduce the continuous use of medicine that has internal absorption organic emulsifiable. In the early stage of mango growth, the growth regulator can be used to promote flowering uniformly to facilitate orderly flowering. However, attention should be paid to the dosage and frequency of the regulator, so as to avoid the persistent flowering spikes and clusters that can not fall off after flower fading to breed a large number of dew spot pathogens; In the middle and late stage of mango growth, the use of growth regulators such as thifenuron and ethephon should be reduced or used at intervals as much as possible, especially not excessive, so as to avoid damaging the peel and the powder of fruit, and reduce the forming conditions of dew spots.

Seventhly, control the disease by chemical. The dosage and frequency of strong permeable pesticides should be reduced as far as possible to avoid the increase of skin permeability when chemicals are used to control mango dew spots. The specific requirements as follows: first, the stone sulfur mixture or Bordeaux liquid should be sprayed to prevent the disease after the mango garden is pruned and before the mango blossom. Second, carbendazim, chlorothalonil and other absorbable fungicides can be applied at the early fruiting stage for protection. Third, monitor and control well in the late fruiting stage. Strengthen the monitoring especially in heavy fog, heavy dew, rain and other special weather to observe the situation of fruit injury. The chemicals should be applied immediately to control when the incidence rate reaches 3% in the field.

4.3 Mango harvesting and fresh-keeping technologies

4.3.1 Harvesting

4.3.1.1 The methods of evaluating the maturity

(1) From appearance

① The fruit has stopped growing and reaching the average weight of the variety; ② The color of pericarp changes from green to apple green, yellow green or light green, or some varieties have white wax layer or micropyle; ③ Mature fruit is found on the tree, or fruit flies and moths injure the fruit; ④ The shape of the fruit shoulder changes. The mature fruit shoulder is full and round, which has the characteristics of this variety; ⑤ The firmness of the pulp is measured with a fruit firmometer, generally 0.172–0.196 MPa; ⑥ The maturity was judged from the flesh color, which changes from milky white to white yellow to light yellow, and from the juice of the cutting fruit,which is white and sticky, reaches the harvest.

(2) Fruit growth days

Generally, it is calculated from the full flowering period, and each variety has a certain number of days. The development days of varieties in different maturity periods are different, such as the development days of early maturing variety Yuexi 1 are 85–100 d, and purple flowers and sweet scented osmanthus mango are 109–125 d. High temperature and drought can lead to early maturity, while high humidity or more rain can delay maturity.

(3) The measurement of contents

When the pH value of fruit is generally above 3.2 and the soluble solids are generally above

4%–6%, it is mature. There is a large difference between different varieties. For example, when the total solid content of Lvsongmang reaches 6.5%, and the maximum value of citric acid is 2.5%, it reaches the light ripening standard.

In addition, there is the gravity method, that is, the proportion of mature mangoes is greater than 1, and those that sink or semi sink in water reach the harvest maturity.

4.3.1.2 Harvesting technologies

In the process of harvesting, all the mechanical damages, such as bruises, scratches, falls, etc., should be prevented, and the gumming pollution from the fruit handle should be avoided as far as possible not to damage the peel. The operation should be strictly performed in accordance with the technical requirements, and the fruit harvesting should be done without damage. Please pay attention to the following things when harvesting.

First, the harvesting time should be selected at 9:00 after the dew dries or at 16:00-18:00 when the sun is not exposed. At this time, the fruit handle has less gum to reduce the pollution of the fruit surface. In addition, it is not suitable to pick fruits in rainy days to stop bacterial infection.

Second, the method of "one fruit, two cuts" should be adopted when picking fruits. That is, cut the whole ear or single fruit from the tree with a cutter, put it in the plastic baskets and then cut it at the 0.5 cm length of the fruit handle after the baskets are transported indoors, so as to prevent the fruit surface from being polluted by gum flowing.

Third, the fruits should be handled with care after harvest. They should not be stacked in the blazing sun, nor directly stacked on the mud and cement ground. The fruits should not touch the ground after being harvested from the trees.

4.3.1.3 Fruit processing, grading and packaging

(1) Processing

The milk and mud on the peel should be washed with water or 1% acetic acid within 8 hours after harvesting, or it can be treated with hot medicine bath (see storage and preservation).

(2) Grading

It should be graded according to the requirements of commodity standards. The first grade fruit is of correct shape, consistent shape and size, smooth skin, and free of disease spots, insect bites and other injuries; The secondary fruits are not seriously damaged (only a few fruits have slight scars). All fruits damaged by fruit flies, fruit sucking armyworms, stalk rot, or fruits with many anthrax spots cannot be used as commercial fruits. In addition, it can also be graded according to the fruit size of the different varieties. The fruits of same size and same varieties and the varieties are in the same boxes.

(3) Packaging

After washed or treated by the hot water and hot bath, the fruit is dried and cooled, and single packaged with clean, soft white paper or 0.11 cm thick polyethylene film, and then the fruit is placed with the pedicel downward and the convex and concave side upward to prevent the juice from flowing out of the fruit to damage the appearance after packaging. Put the fruits into the corrugated cardboard boxes that can carry different weights.

4.3.2 Storage and fresh-keeping

4.3.2.1 The fresh-keeping after postharvest

(1) Anti-corrosion treatment

It is one of the measures that must be taken for commercial packaging of mangoes to soak the fruits in hot water after harvest. Soaking the fruits in hot water of 55 °C for 5 min, of 50–55 °C for 15 min, of 47 °C for 20 min can effectively prevent the damage from anthrax during storage and transportation. Adding fungicides such as benzendazim and thiram in hot water can achieve better results.

(2) Coating

The coating can reduce the water loss and weight loss of fruits respiration and metabolism during storage and transportation. It can also inhibit microbial invasion and growth. The mango can be coated with 6% dewaxed lac solution (containing 0.25% biphenyl).

(3) Chemical treatment

Chemical treatment can prolong the life of mango. The method is using hormones to extend the ripening time of the fruit, and soaking to increase the hardness of the fruit, so as to extend the storage time of green ripe fruit. The fruit, of which the ripening was obviously delayed and the weight loss was less, should be treated with 0.1% surfactant Tripol (cationic detergent) and 100–200 mg/L GA solution at 28–37 °C. Furthermore, after being treated with 32 mg ethylene oxide and stored at room temperature for 16 days, the peel is golden yellow with good flavor. In addition, the treatment of mango with 1-methylcyclopropene, calcium and other chemicals has great effect to reduce the physiological weight loss during the storage period, to delay the ripening, and to maintain the fruit hardness.

4.3.2.2 Packaging

Generally, clean and soft white paper or 0.1–0.2 mm thick polyethylene film is used to package single fruit as the inner packaging, and then they are put into different corrugated boxes that can hold 5 kg, 10 kg or 20 kg weight. The box is divided into two layers: Each layer is separated by cardboard, and each layer is divided into 20–30 grids. Each grid contains one mango. The grid size should be consistent with the fruit size. Then, a small trademark with obvious color and simple design should be pasted on the fruit to beautify and promote the product.

4.3.2.3 Storage

(1) Storing at low temperature

Mangoes should be stored at low temperature for long-term storage or transportation. Proper low temperature can prolong the storage of mangoes. The optimal storage temperature of mangoes is 9–13 ℃, and the relative humidity is 85%–90%. However, there are differences between varieties of mangoes as well as different treatment methods.

(2) Storing in controlled atmosphere

Storage in controlled atmosphere is used to adjust the proportion of oxygen and carbon dioxide concentration during mango storage, and it is effective to delay the respiratory peak. There are combustion type, cracked ammonia type, molecular sieve adsorption type and other types of controlled atmosphere machines. Among them, coke molecular sieve gas conditioner is more practical. It has multiple functions of reducing oxygen, carbon dioxide and ethylene.

(3) Storing at normal temperature

The storage at normal temperature is generally used for short distance transportation and short storage time. Soak the fruit within 500 mg/kg carbendazim in hot water, dry it in the air and bag it. The storage and fresh-keeping period can reach 10–15 d at room temperature (about 30 ℃). Generally speaking, the storage period at room temperature cannot exceed 15 d, otherwise the quality will decline and the flavor will fade. When storing at room temperature, it must be noted that the packing box should be left with basket space to facilitate ventilation and heat dissipation.

4.3.2.4 Acceleration of ripening

Generally, artificial ripening is used to promote ripening in commercial. The mango ripening days can be 4–6 d if the fruit is directly soaked within 500–2,000 mg/kg ethephon for 5 min. In addition, ethephon and abscisic acid can be used to accelerate the ripening of mangoes, and lead to the increase of soluble sugar, so as to improve the quality. Furthermore, the amount of calcium carbide (aluminum carbide) should be 1/2,000–1/100 of the fruit weight. Put mangoes into the low box as sealed as possible without air leakage. Place calcium carbide wrapped in paper or cloth bags at the bottom of the fruit. Calcium carbide will produce ethylene gas after absorbing moisture atmosphere in 48 hours of sealing, so as to accelerate the ripening of mangoes.

4.3.3 Mango products

4.3.3.1 Mango puree

The harvest period of mango is short, and it is not easy to store for a long time. A large number of surplus mature mangoes are often processed into mango puree to store as soon as possible in the place where they are rich, so as to provide raw materials for processing mango drinks, mango jam, mango wine and other mango products in the future. One of the most effective and the fast preservation ways to store mango with high quality commonly used in the world is the rapid processing of mango. The traditional producing

Figure 4-12　Mango puree

process of mango puree is: the mature mangoes → cleaning → selecting → beating → protecting the color→grinding the glue → filtering → blending → degassing → sterilizing → packaging → cooling →storing the finished products. The preservation period of mango can extend from less than 50 days to more than one year by processing into mango puree.

4.3.3.2 Dried mango

It is the most commonly used technology in mango processing, the main process is: selecting the raw material → cleaning → peeling and slicing →protecting the color → drying → softening and packaging.

(1) Selecting the raw material

Choose the fresh and plump fruits that are free from rot, pests or mechanical injuries. The varieties with high dry content, thick and tender flesh, less fiber, small and flat nucleus, in bright yellow color and good flavor shall be selected. The most suitable fruit is 89% of the maturity. If the mango is too green, the color and flavor of mango will be worse. If the mango is too ripe, it will be easy to rot.

(2) Cleaning

Pour the mangoes into the tank with flowing clear water and clean them one by one, furthermore, remove the unqualified fruits, and finally put them into the plastic basket according to the different size and grade, and drain the water.

(3) Peeling and slicing

Use a stainless steel knife to cut off the peel and remove the scar manually. The surface shall be cut smoothly without obvious sharp corners. The outer peel must be removed entirely because the peel contains more tannins. If the peel is not cut cleanly enough, it is easy to browning and affect the color of the finished product during the processing. The peeled fruit is sliced longitudinally with a sharp blade into 8–10 mm thickness pieces and the left pit with residual pulp can be sent for juice making. The technology and process of dried mango.

(4) Protecting the color

The sliced flesh should be immediately put into the color protecting liquid to hold the color and the flavor as well as to stop the rot.

(5) Drying

The raw materials after color protection treatment shall be evenly placed on the bamboo screen (the flesh shall be drained before soaked in the sulfur), and then put into the dryer to dry. The temperature shall be kept at 70–75 °C in the initial stage of drying and down to 60–65 °C in the later stage. During the drying process, pay attention to change the screen, to turn over, to

Figure 4-13　Preparation process and finished product of dried mango

return the moisture and other operations.

(6) Softening and packaging

When the dried mango are of 15%–18% moisture, they should be put into a closed container to soften for 2–3 d, in order to reach a balanced moisture and soft texture for each part, which is convenient for packaging.

4.3.3.3　New health drink of mango

Mango beverage is the main product of mango processing. At present, it mainly includes mango milk beverage, compound beverage, pulp beverage, etc.

Figure 4-14　Mango drink

4.3.3.4　Mango powder

The mango powder is a natural fruit powder refined from fresh mango by vacuum freeze-drying. It is fine, smooth and easy to dissolve, which maintaining the original flavor and nutrients of mango, and has broad prospects for market development.

Figure 4-15   Mango powder

4.3.3.5   Mango jam

Mango jam is a kind of food that is suitable for all ages. It has the unique flavor, good appearance and fine preservation. It can not only be directly used for spreading on the steamed bread or bread in family kitchens or restaurants, but also used as filling in the cakes and pastries. It has great competitiveness in the market.

Figure 4-16   Mango jam and mango smoothie

Current Situation and Future Development of China's Mango Industry

5.1 Current situation of China's mango industry

Among more than 100 mango producing countries and regions, China is the second largest country with abundant variety resources and wide producing areas. In 2022, the mango planting area has reached 349,400 hm^2, the total output is 3.306 million t, accounting for 8.75% of the global output, and its production value achieved 20.52 billion yuan. Meanwhile, China enjoys the reputation of the top import and export country, with surplus of 9.432 million dollars in 2020.

In recent years, progress has been made in scientific research of mango. A cross-regional, interdisciplinary, cross-department and widely-covered whole industrial chain for scientific innovation and breakthroughs system has initially established. The Mango germplasm resource nursery of the Ministry of Agriculture and Rural Affairs has more than 400 preserved resources and has taken its place in the front of the world in completin the whole genome sequencing of mango. A number of excellent new varieties such as Jinhuang, Guifei, Hongyu, Guiqi and Repin series have been selected and bred, and regional distribution system of early, middle and late maturity has been established to realize the annual supply of fresh mango fruits.

5.2 Development of China's mango industry

5.2.1 Mango industry development in recent ten years

Until 2020, mango orchards in China have reached 349,400 hm², and the rapid expansion of mango industry is reflected in the rapid development of output, output value and planting area. From the year 2011 to 2020, the planting area of mango have increased annually from 140,200 hm² to 349,400 hm². The increment is 209,200 hm² with an increase rate of 149.26% and the annual compound growth rate is about 10.68% (Figure 5-1).

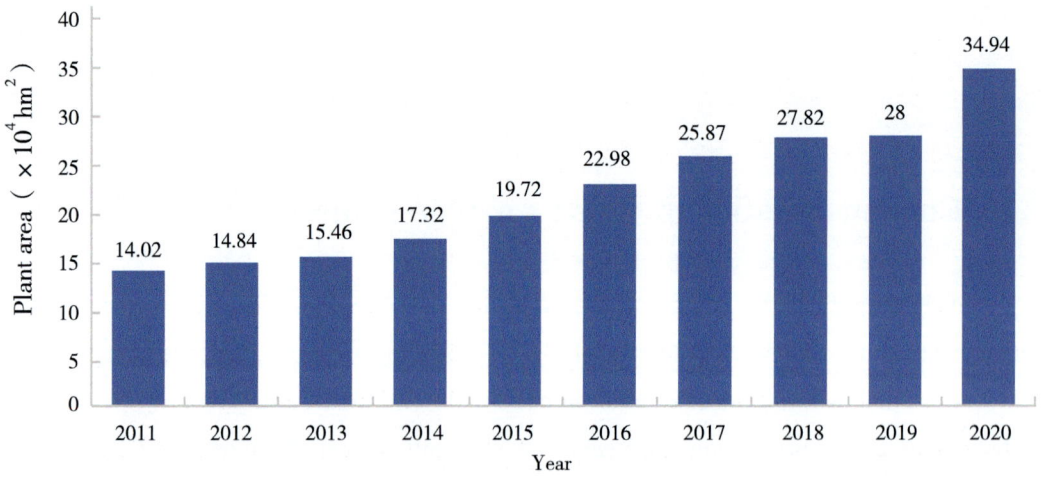

Figure 5-1　Planting area of mango in China from 2011–2020

In recent 10 years, the per unit yield of mango has increased from 7.16 t/hm² to 9.46 t/hm² in 2020, with the increment of 2.30 t/hm² and the annual compound growth rate is about 3.14% (Figure 5-2).

5.2.2 Mango trade in China

At present, countries that have obtained mango access qualification in China market mainly include Australia, Pakistan, Peru, Vietnam, Myanmar, Philippines, Thailand, Ecuador, India, and etc. With the changes of access policy, if other mango-producing countries obtain access qualification in Chinese market, the competition in Chinese mango market will be more fierce.

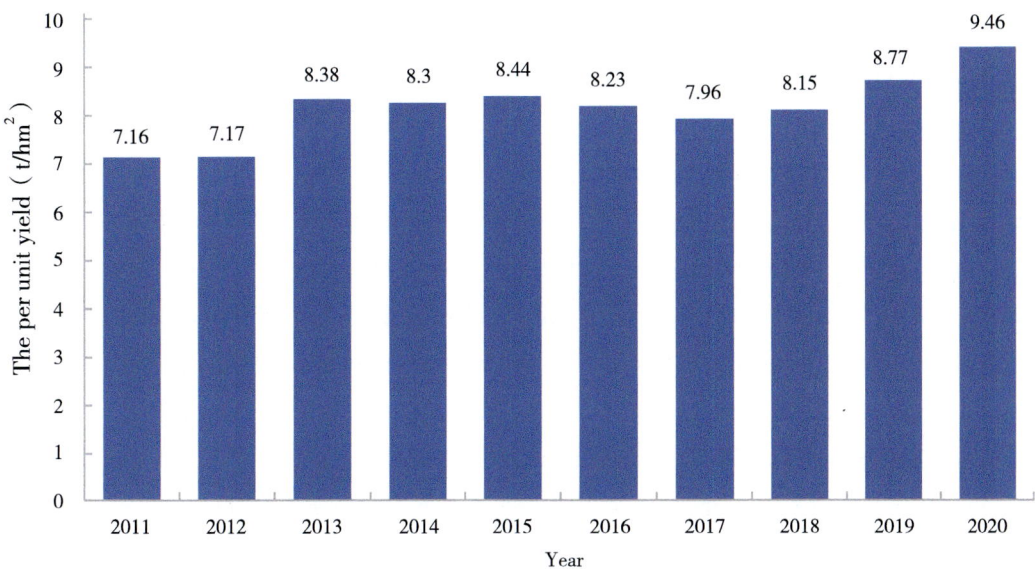

Figure 5-2　Per unit yield of mango in China from 2011–2020

Domestic mangoes will take a bigger hit from imported ones.

With the close economic and trade exchanges between China and ASEAN, mango, as the dominant competitive product in ASEAN countries, has been exported to China in large quantities in recent years, which has caused a great impact on the mango industry in China (Figure 5-3).

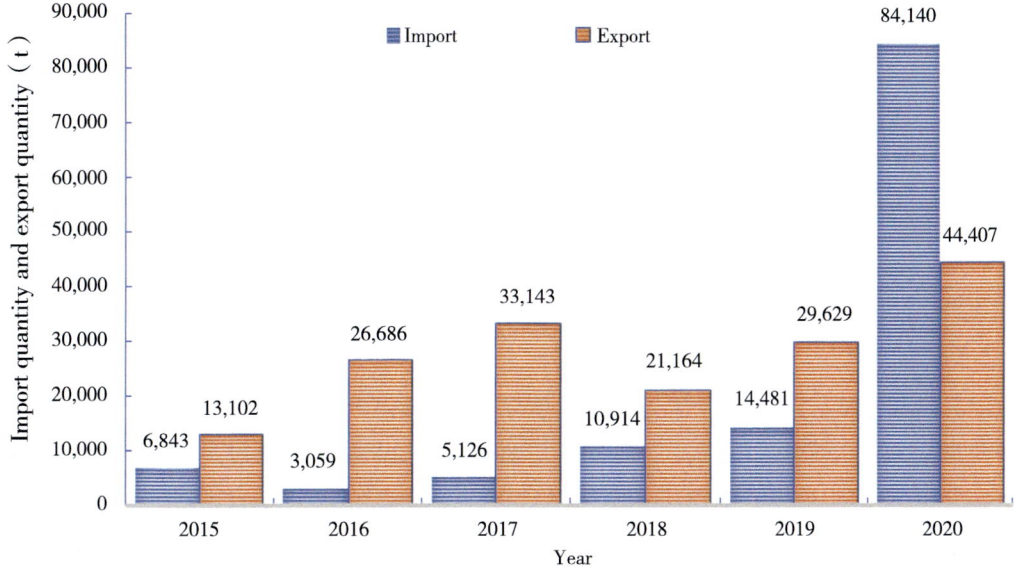

Figure 5-3　Statistics of import and export mangoes in China from 2015–2020
[source: The General Administration of Customs of China (GACC)]

From the structure of import and export statistics, China imported mangoes mainly from Vietnam and Thailand, with Vietnam mangoes accounting for 79.84% of the total imported volume. The Vietnam mango has the characteristics of long supplying period and sufficient supply. Its supplying period lasts from September to the next year's June, and has comparative partial price advantage compared with domestic mangoes. From the perspective of retail markets in Guangdong Province, Guangxi Province and other regions, Vietnam mangoes are mainly Daqing mangoes, whose price is slightly lower than that of Hainan and have strong competitiveness. In addition, the Australia mango has better appearance, higher price, targeted at the mid-range market; while Mangoes from Thailand, such as Rainbow mangoes, are expensive and targeted at the higher end of the consumer market.

Statistics from Custom showed in 2022, China's mango export volume reached 44,407 t, and valued 85.7 million dollars, and main export countries were Vietnam, Russia, Malaysia and etc. Compared with that in 2011, whose export volume was 3,867 t and export value was 2.82 million dollars, the export volume increased by 40,540 t and achieved 10.5 times increase, with the annual compound growth rate of 31.16%; the export value increased by 82.22 million dollars and achieved 29.4 times increase, with the annual compound growth rate of 46.14%. With the assignment of RCPE Agreement and the influence of the pandemic, the export volume increased by 84,140 t and the corresponding period export increased by 44,407 t.

From 2016 to 2020, the exported mango products were almost primary commodity, including fresh and dry mangoes, which were also the main imported products. It is worth noting that the overall trade surplus of this category has been on a downward trend, falling below ten million dollars to 9.423 million dollars in 2020. The main processed product is mango juice. Although the export proportion is relatively small, the export volume increased from 65,000 to 166,000 dollars, with an average annual growth rate of 26.2%, more than twice the average annual growth rate of fresh and dry mango. With the improvement of processing technology and capability of Chinese enterprises, the export proportion of mango processed products is expected to further increase.

At the same time, China also imported mangoes from countries including Thailand, Vietnam, Australia, and Peru, From 2016 to 2020, imports from Vietnam and Thailand all showed an increasing trend, with the Vietnam being particularly significant, while imports from Peru and Australia decreased slightly. Vietnam is both the largest export destination and the largest importer of Chinese mangoes. In 2020, China imported 67,200 t of mangoes from Vietnam, accounting for 79.9% of the country's total import volume. The imported volume was

48,640,300 dollars, accounting for 63.8% of the country's total imports. It is worth mentioning that after Cambodian mangoes were granted access into Chinese market, and the first batch of Cambodian mangoes have arrived in the Chinese market in May, 2020, and the import volume has reached 211 t in four months.

In contrast to the import and export of fresh mangoes, China imports far more mango juice than it exports. There is a huge vacancy in China's mango processing industry. In 2019, the export volume was 221.5 t and the import volume of mango juice in China was 1,805.7 t, which is 8 times of the export volume.

Mango plantation is mainly distributed in Hainan, Guangdong, Guangxi, Fujian, Yunnan, Sichuan, Guizhou and other regions. In 2020, China's mango planting area is 349,000 hm^2, with a total output of 3.306 million t and an output value of 20.52 billion yuan. According to statistics, Hainan and Guangxi account for more than 50% of the mango output in China. Yunnan and Fujian are the provinces with the largest mango export volume. In 2020, Yunnan exported 33,000 t of mango, accounting for 73.9% of the country's total export volume. Exports amounted to 64,215,000 dollars, accounting for 74.9 percent of the country's total exports.

5.3 Current problems and risks in China's mango industry

5.3.1 Current problems in China's mango industry

One is rising labor costs. China's mango industry is generally labor-intensive and technology-intensive work, including the management process of pruning, spraying, planting, fertilizing, fruit picking, fruit and fruit period, etc., with high labor intensity. At present, the labor force accounts for about 50% in the production, and the labor cost in the production keeps rising.

Second, the industry is growing too fast, and regional competition and sales pressure are intensifying. In recent years, some production areas have increased by more than 70%, and the production capacity has increased rapidly. Under the double impact of domestic fruit overcapacity and the import competition of fruits from South America and Southeast Asia, low prices and difficult selling has hindered the development of mango market if there is a lack of effective sales channels, and abundant production could not bring high incomes.

Third is the low degree of organization, and the serious homogeneous competition of all links in the industrial chain, leading to the decline of the overall competitiveness of the industry.

At present, farmers are main body of the mango management and it is difficult to unify the production technology, standardize the production process; thus the quality of the fruit produced is mixed, the number of leading enterprises is limited, the capacity of storage, preservation and transportation technology is insufficient, deep processing and post-harvest treatment technology is extensive, and all these factors affect the economic benefits. At the same time, a large number of fruit farmers are not only producers, but also take the role of technicians and salespeople. The division of labor is not clear, which seriously restricts the healthy development of the industry.

Fourth, mango's seasoning differentiation advantage weakened. China is a rare country that could make use of regional characteristics and technological advantages to realize the annual seasoning of mango fruits. However, with the development of science and technology, a large number of countries in ASEAN and South Asia can also realize the adjustment of mango production period. The differentiation advantage in producing areas of Chinese mango weakens and profits decrease.

Fifth, low degree of information technology and weak construction of infrastructure. Information flow is not smooth, market information is insufficient, farmers can not timely adjust the product structure, unable to meet the needs of the market, which is not conducive to market development. Fruit farmers' investment of orchard is insufficient, new technology, new products can not be widely promoted and applied, low-yield orchard area is large with low economic benefits.

Sixth, lack of innovation and investment. Mango scientific research focuses on pre-production and mid-production, while more efforts should be paid on post-production. The mango industry lacks standardization, industrial innovation research and support investment. At present, the mango market access system has not been established, the market regulation function is lacking, far from the standardized production system.

5.3.2　Risks in China's mango industry

One is the excessive use of hormones, by forcing flowers to produce early fruits. In previous years, many mango growers forced the early harvest of mango, and sold high prices in the domestic mango production gap period around the Spring Festival. The planting efficiency formed a certain driving effect. However, spraying unqualified hormone to leaf surface or small fruit or even directly soaking fruits in hormone violated the law of natural growth, causing many problems such as mango quality decline, food safety and frequent disease.

Second, mango industry chain is short, lacking deep processing. At present, there are only more than 10 large tropical fruit processing plants in China, and the deep processing capacity is obviously insufficient. In addition to the parts with good fruit quality and easy to sell, the remaining 30%-40% secondary fruit needs to be purchased by the processing plant in the mango harvest season every year, and the price is as low as 0.4 yuan /kg, which is difficult to sell or dispose. The lack of mango deep processing links restricts the mango industrial chain, the added value can not be improved, resources can not be rationally allocated, greatly limiting the mango industry to play its due benefits.

Third, post-harvest preservation technology has not been widely used. In Hainan, mangoes are generally not preserved after harvest. They are usually picked, graded and packaged at the origin and directly sent to wholesale markets outside the province by buyers. In the past few years, mangoes were mostly picked in a state of immaturity, stronger during the transportation; thus these fruits could enter into the market earlier, and stay fresh in the wholesale market longer. Now the government is encouraging farmers not to pick the tender fruit, which not only affects the taste of the mango, but even inedible. In fact, the heat treatment technology of mango's post-harvest preservation has been relatively mature in the field of scientific research; however, to save cost, this technique is rarely known or used by farmers. We should strengthen farmers' awareness of post-harvest preservation, improve the technical level through farmer training, so that mango can be preserved and transported scientifically.

5.4 Mango market and industry prospect in China

China's mango market has great potential, but domestic circulation and channels still need be further expanded. Guangxi, Yunnan, Sichuan and Cambodia have a large area of newly planted mango in recent years, and the fruit will be harvested in three to five years, when the domestic mango output will increase rapidly; meanwhile, the newly planted mango of Cambodia by Chinese enterprises has harvested and will directly sell to the domestic market. If the circulation and marketing channels are insufficient or could not meet the demand of these mangoes, there will be a risk of oversupply of fresh fruit and its price will certainly decline.

The high point of mango price in China is from December to February, and the low point is from May to September. The main reason is that the insufficient supply of fresh fruits in winter, and the strong demand of consumption during holidays (Spring Festival), which leads to the increase of mango wholesale price. However, from May to September, when a large number of fresh fruits, especially lychee, longan, are sold on the market, consumers have more choices, which brings down the wholesale price of mango.

At the same time, transport loss and quality deterioration are also important factors affecting mango value chain. Now only a few enterprises realize the need to use modern logistics to reduce the loss of transport links, such as Sanya tree ripe mango, product price is high, but thin skin, short storage life, can not withstand transport. In 2017, a fruit professional cooperative in Sanya sold more than 1 million kilograms of mango ripe on trees to the outside of the island through the whole cold chain logistics through the fresh e-commerce platform. The cooperative specially purchased a small cold chain transport vehicle in the field. After the mango is picked, it is immediately moved onto the cold chain truck, pulled to the air conditioning and refrigeration of the constant temperature packaging workshop. After being packed, it is directly transported by cold chain logistics to the cold storage or consumers in the local market. In addition to increasing the cold chain transportation cost, the whole cold chain logistics also adds an additional handling cost. Although it is time-consuming and costly, only a good cold chain transportation can improve the planting income.

Production and consumption expectations. Domestic market is not saturated and demand of high quality mango is strong. In recent years, different traditional consumption of fresh mangoes, the market share of mango juice, dried mango and mango paste has increased rapidly. The high price and strong demand have encouraged more farmers to plant mango. Considering the processing cost, mango processing plant needs to import fresh fruit or raw pulp from abroad to supplement the raw materials. If no abnormal climate happens in the future, mango production and consumption will show a good dual-track development.

Market prospect analysis. Driven by domestic consumption, mango products are performing well in wholesale and retail markets throughout the country. High-quality mangoes, imported from Australia, Thailand, Peru and Ecuador, etc, are mainly sold in large chain supermarkets in first-tier cities such as Beijing, Shanghai, Guangzhou and Shenzhen. The average price of these mangoes is about 60 yuan/kg, much higher than that of domestic mangoes. Australian mangoes are the most expensive in high-end supermarkets, with the price around 200 yuan/g, which indicates that Chinese market has the strong demands for high-quality mangoes.

5.5 Outlook

With the development of economy and the improvement of consumption level, people's consumption demand for mango is also growing, and mango production has become a pillar industry in the development of tropical and subtropical rural economy. However, with the development of science and technology, changes of production and management environment, the traditional mango planting has been strongly impacted. In the new era, the healthy and sustainable development of mango industry depends on high quality, cost-saving, environmental friendly, safety, high efficiency of modern cultivation, post-harvest preservation and processing technology. At present, there are still some problems in our mango industry, such as slow optimization of variety structure, short industrial chain, low added value, and weak brand effect. In the next step, we should make efforts in promoting the mango industry, market and cultural values, and rely on science and technology to achieve the sustainable development of mango industry.

First, optimize mango varieties and improve its quality. It is necessary to take scientific planning as the guide, take the holistic approach in the industrial layout, and strengthen the construction of mango germplasm resource nursery and innovation base through the integration of scientific and technological resources, accelerate the discovery of a number of excellent germplasm, cultivate superior varieties and breeds with high yield and quality, build a superior seed-breeding base, and improve the capacity of mango seed supply.

Second, strengthen the foundation of industrial development and extend the industrial chain. It is necessary to gather land, capital, talents, information and other factors, improve the level of industrial organization based on base construction, and create the whole industrial chain development mode of mango "R&D + production + processing + sales + tourism". We will accelerate the construction of orchards, storage, logistics and other infrastructure, ensure the capacity of sorting, processing, preservation, storage and transportation, and realize efficient and low-loss circulation from the field to the table. To increase the deep processing of mango, improve the added value, enhance the risk defense ability of the industry.

Third, expand sales channels and build competitive brands. Improve the network sales system, promote cooperation with international purchasers, and cultivate mango export demonstration bases and export enterprises. We will accelerate the cultivation of dominant brands and strive to cultivate a number of leading domestic and internationally renowned regional public brands, enterprise brands and product brands.

Fourth, cultivate mango culture and promote industrial upgrading. It is necessary to strengthen the cultivation of local technical talents and socialized technical services, explore the integration of mango industry development with clean water and green mountains and leisure agriculture, activate the endogenous power of rural revitalization, promote production standardization with industrial organization, take brand building as the strategy of industrial development, constantly improve the quality, efficiency and competitiveness of mango industry, and lead the green and healthy development of the industry.

参考文献

冯春梅,李建强,温立香,等,2015.原味芒果干无硫加工品质提升的工艺优化[J].南方农业学报,46(7):1292-1296.

高爱平,陈业渊,朱敏,等,2006.中国芒果科研进展综述[J].中国热带农业(6):21-23.

黄国弟,苏美花,王春田,2010.我国芒果标准化生产现状及发展对策[J].中国热带农业(2):4.

黄艳,2018.巴基斯坦芒果研究所引进高密度芒果种植园[J].世界热带农业信息(10):55.

霍建华,苏桂新,2017.试论芒果高产种植技术的要点[J].现代园艺(13):68.

柯佑鹏,黄良团,2013.2012年海南芒果产业损害监测预警分析报告[J].中国热带农业(2):40-43.

赖必辉,毕金峰,庞杰,等,2011.芒果加工技术研究进展[J].食品与机械(3):4.

李华丽,魏仲珊,陈瑶,等,2012.芒果及其加工制品研究进展[J].农产品加工(学刊)(10):113-116.

李丽,盛金凤,孙健,等,2014.芒果加工新技术及综合利用研究进展[J].食品工业,35(6):223-227.

李日旺,黄国弟,苏美花,等,2013.我国芒果产业现状与发展策略[J].南方农业学报,44(5):4.

李晓娜,曾小红,谢龙莲,等,2017.世界芒果炭疽病防治技术研究概况[J].热带农业科学,37(11):7.

卢桂仙,李树榜,2019.对山地芒果高产优质种植技术的思考[J].种子科技,37(4):76.

罗学兵,2011.芒果的营养价值、保健功能及食用方法[J].中国食物与营养,17(7):77-79.

尼章光,陈于福,解德宏,等,2013.云南芒果产业发展规划研究[J].中国农业资源与区划(3):6.

彭杨,黄海,龚德勇,等,2021.栽培技术对芒果果实品质影响的研究进展[J].农技服务(7):56-58,62.

汪汇源,2016.新芒果品种在印度普及[J].世界热带农业信息(7):9.

臧小平, 周兆禧, 林兴娥, 等, 2016. 不同用量有机肥对芒果果实品质及土壤肥力的影响 [J]. 中国土壤与肥料（1）: 4.

张劲, 黄丽, 夏宁, 等, 2011. 6个芒果品种品质特性评价研究 [J]. 食品科技, 36（9）: 5.

张鲁斌, 常金梅, 詹儒林, 2010. 芒果采后病害生物防治研究进展 [J]. 热带作物学报（8）: 6.

郑素芳, 张岳恒, 2011. 海南芒果产业链现状研究 [J]. 中国农业资源与区划, 32（2）: 6.

钟丽琪, 王晓雯, 郑云芳, 2017. 芒果加工的综合利用综述 [J]. 现代食品（12）: 65-66.

钟勇, 黄建峰, 罗睿雄, 2016. 海南省芒果产业化发展现状、存在问题及对策 [J]. 中国热带农业（3）: 19-22.

SANTOS L, LIMA A, CUNHA J C, et al., 2019. Does irrigated mango cultivation alter organic carbon stocks under fragile soils in semiarid climate? [J]. Scientia Horticulturae, 255: 121-127.

CHAUDHARY M V, KHODIFAD P B, THAKUR N B, 2020. Awareness of mango growers about good agriculture practices in mango cultivation[J]. Journal of Global Communication, 13（1）: 10.

SMITH H R, 2020. The Mango: Botany, Cultivation, and Utilization[J]. Journal of Association of Official Agricultural Chemists, 45（3）: 794.

YADAV R N, SHARMA T D, 2007. Knowledge and adoption level of orchardists regarding mango cultivation in Western Uttar Pradesh[J]. Plant Archives, 7（1）: 389-393.